Olfa Kanoun, Faouzi Derbel and Nabil Derbel (Eds.)
Sensors, Circuits & Instrumentation Systems

Advances in Systems, Signals and Devices

—

Edited by
Olfa Kanoun, University of Chemnitz, Germany

Volume 6

Sensors, Circuits & Instrumentation Systems

Extended Papers 2017

Edited by
Olfa Kanoun, Faouzi Derbel and Nabil Derbel

DE GRUYTER
OLDENBOURG

Editors of this Volume

Prof. Dr.-Ing. Olfa Kanoun
Technische Universität Chemnitz
Chair of Measurement and Sensor Technology
Reichenhainer Strasse 70
09126 Chemnitz, Germany
olfa.kanoun@etit.tu-chemnitz.de

Prof. Dr.-Eng. Nabil Derbel
University of Sfax
Sfax National Engineering School
Control & Energy Management Laboratory
1173 BP, 3038 SFAX, Tunisia
n.derbel@enis.rnu.tn

Prof. Dr.-Ing. Faouzi Derbel
Leipzig University of Applied Sciences
Chair of Smart Diagnostic and Online Monitoring
Wächterstrasse 13
04107 Leipzig, Germany
faouzi.derbel@htwk-leipzig.de

ISBN 978-3-11-044619-7
e-ISBN (PDF) 978-3-11-044837-5
e-ISBN (EPUB) 978-3-11-044624-1
ISSN 2365-7493

Library of Congress Control Number: 2018934304

Bibliographic information published by the Deutsche Nationalbibliothek
The Deutsche Nationalbibliothek lists this publication in the Deutsche Nationalbibliografie;
detailed bibliographic data are available on the Internet at http://dnb.dnb.de.

© 2018 Walter de Gruyter GmbH, Berlin/Boston
Typesetting: Konvertus, Haarlem
Printing and binding: CPI books GmbH, Leck

www.degruyter.com

Preface of the Editors

The sixth volume of the series "Advances in Systems, Signals and Devices" (**ASSD**), is a peer reviewed international scientific volume devoted to the field of sensors, circuits and instrumentation systems. The scope of the volume encompasses all aspects of research, development and applications of the science and technology in these fields. The topics include Sensors and measurement systems, optical sensors, chemical sensors, mechanical sensors, inductive sensors, capacitive sensors, micro-sensors, thermal sensors, biomedical and environmental sensors, fexible sensors, nano sensors, micro electronic systems, nano systems and nano technology, sensor signal processing, sensor interfaces, modeling, data acquisition, multi sensor data fusion, distributed measurements, device characterization and modeling, custom and semicustom circuits, analog circuit design, low-voltage, low-power VLSI design, circuit test, packaging and reliability, impedance spectroscopy, wireless sensors, wireless interfaces, wireless sensor networks, energy harvesting, circuits and systems design.

Every issue is edited by a special editorial board including renowned scientist from all over the world. Authors are encouraged to submit novel contributions which include results of research or experimental work discussing new developments in the field of sensors, circuits and instrumentation systems. The series can be also addressed for editing special issues for novel developments in specific fields. Guest editors are encouraged to make proposals to the editor in chief of the corresponding main field.

The aim of this international series is to promote the international scientific progress in the fields of systems, signals and devices. It is a big pleasure of ours to work together with the international editorial board consisting of renowned scientists in the field of sensors, circuits and instrumentation systems.

Editors-in-Chief
Olfa Kanoun, Faouzi Derbel and Nabil Derbel

De Gruyter Oldenbourg, ASSD – Advances in Systems, Signals and Devices, Volume 6, 2018, p. V.
https://doi.org/10.1515/9783110448375-202

Advances in Systems, Signals and Devices

Series Editor:

Prof. Dr.-Ing. Olfa Kanoun
Technische Universität Chemnitz, Germany.
olfa.kanoun@etit.tu-chemnitz.de

Editors in Chief:

Systems, Automation & Control

Prof. Dr.-Eng. Nabil Derbel
ENIS, University of Sfax, Tunisia
n.derbel@enis.rnu.tn

Power Systems & Smart Energies

Prof. Dr.-Ing. Faouzi Derbel
Leipzig Univ. of Applied Sciences, Germany
faouzi.derbel@htwk-leipzig.de

Communication, Signal Processing & Information Technology

Prof. Dr.-Ing. Faouzi Derbel
Leipzig Univ. of Applied Sciences, Germany
faouzi.derbel@htwk-leipzig.de

Sensors, Circuits & Instrumentation Systems

Prof. Dr.-Ing. Olfa Kanoun
Technische Universität Chemnitz, Germany
olfa.kanoun@etit.tu-chemnitz.de

Communication, Signal Processing & Information Technology

Til Aach, Achen University, Germany
Kasim Al-Aubidy, Philadelphia Univ., Amman, Jordan
Adel Alimi, Engineering School of Sfax, Tunisia
Najoua Benamara, Engineering School of Sousse, Tunisia
Ridha Bouallegue, Engineering School of Sousse, Tunisia
Dominique Dallet, ENSEIRB, Bordeaux, France
Mohamed Deriche, King Fahd University, Saudi Arabia
Khalifa Djemal, Université d'Evry, Val d'Essonne, France
Daniela Dragomirescu, LAAS, CNRS, Toulouse, France
Khalil Drira, LAAS, CNRS, Toulouse, France
Noureddine Ellouze, Engineering School of Tunis, Tunisia
Faouzi Ghorbel, ENSI, Tunis, Tunisia
Karl Holger, University of Paderborn, Germany
Berthold Lankl, University of the Bundeswehr, München, Germany
George Moschytz, ETH Zürich, Switzerland
Radu Popescu-Zeletin, Fraunhofer Inst. Fokus, Berlin, Germany
Basel Solimane, ENST, Bretagne, France
Philippe Vanheeghe, Ecole Centrale de Lille France

Sensors, Circuits & Instrumentation Systems

Ali Boukabache, Univ. Paul, Sabatier, Toulouse, France
Georg Brasseur, Graz University of Technology, Austria
Serge Demidenko, Monash University, Selangor, Malaysia
Gerhard Fischerauer, Universität Bayreuth, Germany
Patrick Garda, Univ. Pierre & Marie Curie, Paris, France
P. M. B. Silva Girão, Inst. Superior Técnico, Lisboa, Portugal
Voicu Groza, University of Ottawa, Ottawa, Canada
Volker Hans, University of Essen, Germany
Aimé Lay Ekuakille, Università degli Studi di Lecce, Italy
Mourad Loulou, Engineering School of Sfax, Tunisia
Mohamed Masmoudi, Engineering School of Sfax, Tunisia
Subha Mukhopadhyay, Massey University Turitea, New Zealand
Fernando Puente León, Technical Univ. of München, Germany
Leonard Reindl, Inst. Mikrosystemtec., Freiburg Germany
Pavel Ripka, Tech. Univ. Praha, Czech Republic
Abdulmotaleb El Saddik, SITE, Univ. Ottawa, Ontario, Canada
Gordon Silverman, Manhattan College Riverdale, NY, USA
Rached Tourki, Faculty of Sciences, Monastir, Tunisia
Bernhard Zagar, Johannes Kepler Univ. of Linz, Austria

Advances in Systems, Signals and Devices

Contents

M. Zabat, N. Ouadahi, A. Ababou, N. Ababou, K. Ben Si Said and
A. Youyou

Digital Inclinometer for Range of Motion Measurements in All Anatomical Planes

Abstract: Range of motion measurement using wearable sensors is an inexpensive, convenient, and efficient manner of providing useful information for joint disorder. Classical digital inclinometers can only measure joint angles in a vertical plane. This paper describes an embedded system that measures in addition the joint angles in a horizontal plane. Its principle of operation is based on gravitational acceleration and magnetic field sensors measurements. The method used to measure the joint angles in the three anatomical planes of a standing or sitting subject for a given position of the device is described. A Graphical User Interface has been developed to display simultaneously the flexion-extension, abduction-adduction and internal-external rotations of the 3-DOF shoulder joint angles during a simple movement performed by a subject. This movement is reconstructed in a 3D real-time animation.

Keywords: Embedded system, range of motion, digital inclinometer, inertial sensor.

1 Introduction

An inclinometer or clinometer is an instrument for measuring tilt angle of an object with respect to horizon or gravity. It can be used in very different applications such as geotechnical instrumentation on transportation projects [1], civil engineering [2] or planet rover localization when combined with a sun sensor [3]. These sensors have also been utilized for measuring cruising yacht free boards [4] as well as in robotics [5] and medical applications [6, 7, 8, 9, 10]. Joint angular sensors are widely used in the industry, from highly effective robots in product lines and heavy construction machines in building sites to small knobs on home appliances [11] as well as in rehabilitation or physical therapy [12, 13, 14]. In medical or clinical applications, the inclinometer is usually used to determine a body's joint range of motion (ROM). Typically, range of motion ROM is referenced from body's natural position. In rehabilitation, if a patient experiences decreased range of motion in a joint, the therapist can use several kinds of goniometers to assess the range of motion. In team sports such as volleyball [15] or handball [16, 17], pain and shoulder instability, as well as scapular dysfunction influence negatively an athlete's performance. In these sports, overhead athlete's shoulders are examined

M. Zabat, N. Ouadahi, A. Ababou, N. Ababou, K. Ben Si Said and A. Youyou: Laboratory of
Instrumentation, Faculty of Electronics & Computer Science, University of Science and Technology
Houari Boumediene, Algiers, Algeria, emails: mzabat@usthb.dz, aababou@usthb.dz

De Gruyter Oldenbourg, ASSD – Advances in Systems, Signals and Devices, Volume 6, 2018, pp. 1–20.
https://doi.org/10.1515/9783110448375-001

with evaluation of the shoulder total range of motion as a sum of the internal and external rotations in 90° arm abduction and 90° elbow flexion while the athlete is in a dorsal decubitus position.The examination of shoulder mobility may be accomplished using visual observation or instruments such as goniometers, electro-goniometers and digital inclinometers [18]. Based on the method proposed above, Goniometry has been used widely due to its portability and low cost. However, a limitation of goniometry is that it requires the clinician to use both hands, making stabilization of the extremity more difficult, and thus increasing the risk of error in reading the instrument. Inclinometry is another practical alternative that incorporates the use of constant gravity as a reference point to assess joint mobility. Digital inclinometers are portable, lightweight, and require training similar to that of goniometry. Digital inclinometer measures provide a quick highly reliable, valid, direct assessment of postural impairments such as trunk extensor muscle weakness or vertebral fracture [9, 19]. To the authors' knowledge, digital inclinometer proposed in literature can only be used to measure flexion-extension angles in the sagittal plane and adduction-abduction in the coronal plane of standing or sitting subject for which these two anatomical planes are vertical. In this case, internal-external rotation in the transverse plane (horizontal plane) cannot be measured by classical digital inclinometer as its principle of operation is based on MEMS accelerometric sensor. Hence, this last joint angle can only be measured with a classical digital inclinometer if the subject is in supine position.

In this paper, a digital inclinometer prototype developed in our laboratory is presented. This embedded system is able to measure simultaneously joint angles of somebody in standing or sitting position in the three anatomical planes. It can be used by therapists and physicians to measure joint angles in a more precise way when compared to the use of classical plastic goniometer which is simply a protractor.

2 Methods

Anatomical planes of a standing body in anatomical position are shown in Fig. 1. A ROM is generally associated with flexion/extension in the sagittal plane (S), adduction/abduction in the frontal or coronal plane (F), and internal/external rotation in the transverse plane (T). So, the ROM of an anatomical joint can be obtained from angles measurement associated with flexion/extension, adduction/abduction and internal/external rotation.

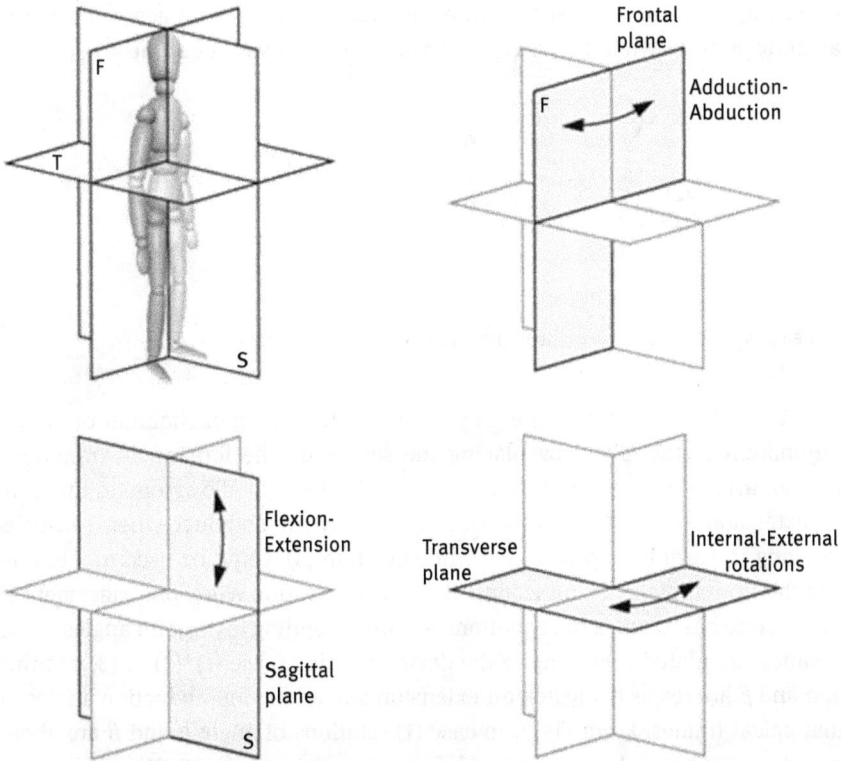

Fig. 1. Anatomical planes of a body in anatomical position.

2.1 Angles determination

The joint angles can be deduced from a three-axis accelerometer and a three-axis magnetometer outputs by considering the set of equations corresponding to a given situation.

2.1.1 Angles deduced from accelerometer

A three-axis accelerometer measures the projection of the gravity vector on its sensing axes. Several methods can be used for measuring tilt angles from a low-g 3-axis accelerometer. The method to be chosen must guarantee a constant sensitivity and improve the accuracy [20]. This can be achieved by using a 3-axis accelerometer to measure flexion/extension in the sagittal plane or adduction/abduction angles in the frontal plane. In order to provide greater flexibility of measurement we proposed a mean that make the device able to detect the sensitive axis to gravity before each measurement according to one of the three possible inclinometer placements on the human body joint shown in Fig. 2. x, y and z are the sensitive axes of the sensor.

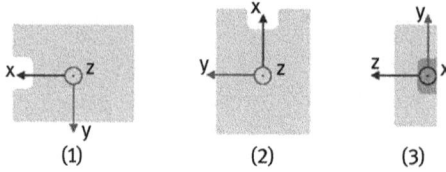

Fig. 2. Three possible placements of the digital inclinometer in a vertical plane.

Position (1) can be used to measure internal/external rotation of shoulder or prona-tion/supination of the elbow by placing the sensor on the forearm in transverse plane when arm is 90°-flexed in sagittal plane. Position (2) allows one to measure flexion/extension and abduction/adduction angle of the shoulder when sensor is placed along the arm in sagittal plane while position (3) would be used to measure angle in the frontal plane as, for example, scapula elevation.When one placement is considered, one has to select the equations set corresponding to the joint angles in the three planes calculated according to the device position (case (1), (2) or (3)). In this paper, α and β are respectively flexion-extension and adduction-abduction angles in the anatomical frames. From Fig. 2, in case (1) rotations of angle α and β are about x-axis and y-axis. α and β are expressed by equations (1) and (2) [20, 21].

$$\alpha = \arctan \frac{A_y}{A_x^2 + A_z^2} \tag{1}$$

$$\beta = \arctan \frac{A_x}{A_y^2 + A_z^2} \tag{2}$$

In case (2) rotations of angle α and β are about z and y. α and β are respectively expressed by:

$$\alpha = \arctan \frac{A_y}{A_x^2 + A_z^2} \tag{3}$$

$$\beta = \arctan \frac{A_z}{A_x^2 + A_x^2} \tag{4}$$

In case (3) rotations of angle α and β are about z and x axes. α and β are respectively expressed by:

$$\alpha = \arctan \frac{A_x}{A_y^2 + A_z^2} \tag{5}$$

$$\beta = \arctan \frac{A_z}{A_x^2 + A_y^2} \tag{6}$$

where A_x, A_y and A_z are the 3-axis accelerometer outputs.

2.1.2 Angles deduced from magnetometer

Magnetometers are sensors that measure the strength and direction of the local magnetic field using two sensitive axes x and y placed in the horizontal plane. The heading γ measured by the magnetometer corresponds to internal-external rotation. It is expressed by:

$$\gamma = \frac{Y_h}{X_h} \tag{7}$$

where X_h and Y_h are magnetic field components in the horizontal plane. Then, equation (7) is valid only when magnetometer is in the horizontal plane. Otherwise, X_h and Y_h need to be corrected using tilt angles α and β measured by the accelerometer to eliminate tilt error. The corrected expressions of X_h and Y_h [21] are shown in the following equations:

$$X_h = X_m \cos \alpha + Z_m \sin \alpha \tag{8}$$
$$Y_h = X_m \sin \beta \sin \alpha + Y_m \cos \beta - Z_m \sin \beta \cos \alpha \tag{9}$$

where X_m, Y_m and Z_m are the magnetometer outputs. X_h and Y_h in equations (8) and (9) were calculated in case (1) of Fig. 2. Angle γ corresponding to internal-external rotation in horizontal plane is expressed by equation (7).

2.2 Block diagram of the inclinometer

Figure 3 shows the electronic system block diagram. The Sensor block is composed of an integrated circuit containing a 3-axis digital accelerometer and a 3-axis digital magnetometer placed in the same package in order to prevent misalignment problems. The accelerometer measures angles in the vertical plane and the magnetometer performs the correct measure in the horizontal plane.

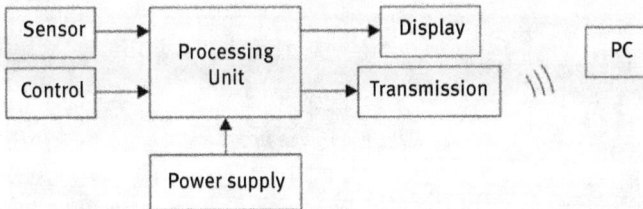

Fig. 3. Block diagram of the developed inclinometer.

The Control block consists of four push buttons (reset, measure, hold and send) each one with a resistance mounted in parallel. The reset button is used to reset the device and the measure button to set the angles to zero before performing the joint angles measurements. The hold button maintains the final value on the display screen and stops the wireless transmission and the send button allows one to select the display mode or the wireless link mode to GUI in the personal computer PC. The processing unit, a PIC18F2550 microcontroller from Microchip performs the following tasks: (i) Data acquisition from sensors via I2C bus, (ii) Calculus of flexion-extension, abduction-adduction, internal-external rotation angles, (iii) Send data to the display, (iv) Send data to PC via a wireless link. The Display block consists of a Nokia 1100 monochrome LCD which has 8x19 alphanumeric characters and 4 cm x3.5 cm x0.3 cm size. The power supply is equal to 3.3 V. Communication between the inclinometer and the PC is achieved via wireless Bluetooth link.In an embedded system, the choice of the power supply is essential. An assessment of the current consummated by the device in work has been found to be less than 100 mA. Using a lightweight and compact lithium-Ion BL-4C battery with a charge capacity equal to 860 mAh can provide autonomy of the system approximately equal to 10 hours.

2.3 Sensors calibration

2.3.1 Accelerometer calibration

Before experimental measurements with the inclinometer, sensors calibration has been carried out. The angles measured by the accelerometer are pitch and roll corresponding to α and β angles (in vertical plane when the accelerometer is in horizontal plane before tilting). As the limb movement is usually performed at low speed, the sensor response time is sufficient for the intended application. The accelerometer calibration has been carried out on the bench realized in our laboratory using a 360° protractor associated with a bubble level and shown in Fig. 4a. The

(a)   (b)   (c)

Fig. 4. (a) Calibration bench; (b) Sensor unit on the horizontal plane; (c) Pitch angle measurements.

calibration consisted in sensor angle tilting from 0 to 180° by 5°-step using the calibration bench as illustrated in Fig. 4b and 4c.

Angles measured in this case were pitch angles. A 90° clockwise rotation of the sensor unit shown in Fig. 4b resulted in roll angles measures.

2.3.2 Magnetometer calibration

The magnetic field measured by the magnetometer is a combination of both the Earth's magnetic field and any magnetic field created by nearby objects. So, in order to be correctly used as compass this sensor needs to be calibrated before its utilization. Magnetometer calibration means elimination of all noisy magnetic fields (hard-iron and soft-iron interference magnetic fields) influencing the hand held device. Hard-iron interference magnetic field is generated by ferromagnetic materials with permanent magnetic fields. These materials could be permanent magnets or magnetized iron or steel. Soft-iron interference magnetic field is generated by the items inside the hand held device. They could be current carrying traces on the PCB or magnetically soft materials. The magnetometer calibration has to be performed each time the device is placed in a different environment in which the magnetic field was modified. Calibration of the magnetometer consists in performing a 360° rotation of the sensor in the horizontal plane and to plot its y-axis magnetic component Y_h versus x-axis magnetic component X_h as the angle varies from to 0° to 360°. The graph representing Y_h versus X_h will be a circle non-centered at the origin if there is only hard-iron effect or an ellipse if soft-iron effect is present, too.

3 Experimental

3.1 Measurement system

A photograph of the digital inclinometer prototype is shown in Fig. 5. It is composed of a main unit wired to the sensor unit by a four-wire phone cord as shown in the center photograph of Fig. 5. A 3.3 V Li-Ion battery is placed in the rear part of the main unit as illustrated in the left photograph, and processing unit, as well as control and display units are involved in the front side. The sensor unit shown in the right picture contains a LSM303DLHC circuit from ST Microelectronics (a system-in-package featuring a 3-axis digital linear acceleration sensor and a 3-axis digital magnetometer).

The sensor unitmust be attached near the joint whose range of motion has to be assessed. Thesensors outputs are related tothe joint angle values as explained in subsection 2.1.1 and 2.1.2.

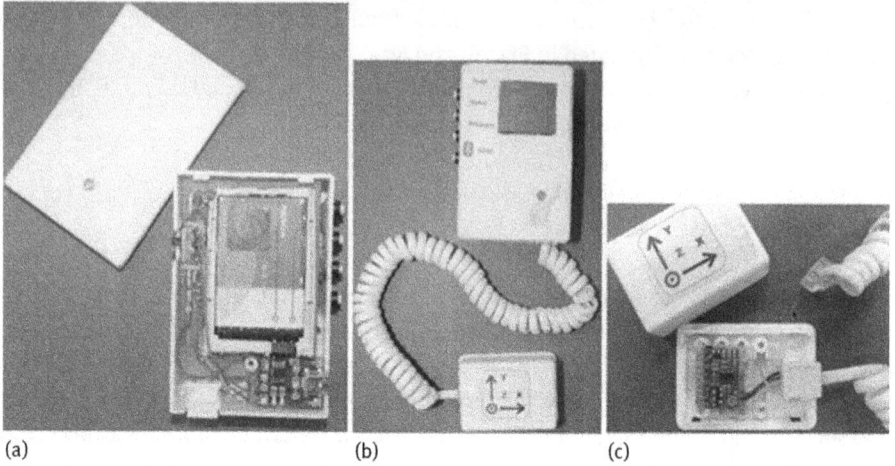

(a)   (b)   (c)

Fig. 5. (a): main unit rear face. (b): main unit wired to sensor unit by a 4-wire cable. (c): sensor unit without cap.

3.2 Software graphical user interface

A software graphical user interface GUI on PC has been developed and implemented to ensure(i) 3D real-time upper limb movement animation, (ii) a graphical display of the measured angles, (iii) generation of a data basis in an Excel file that contains information about the patient as well as the measurements performed by the clinician. Fig. 6 presents the result of the flexion ROM corresponding to the movement performed by the subject shown in Fig. 11.a when he was asked to e maximal shoulder flexion. This GUI was created under Microsoft Visual C# environment with .NET

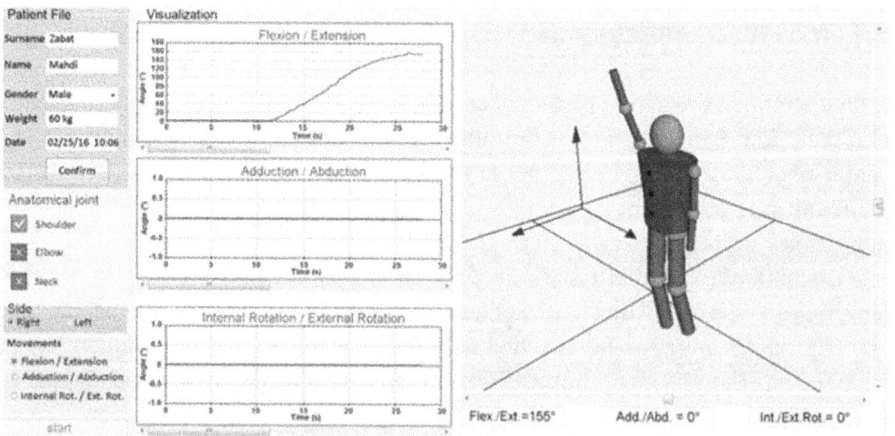

Fig. 6. Graphical user interface associated with shoulder flexion performed by subject in Fig. 11a.

Framework. Concerning the 3D-animation, it is a structure on which the rotation is applied about the three anatomical axes using rotation matrices that are filled with data issued from the device when a subject is performing, for example, a movement of flexion-extension, abduction-adduction, internal-external rotation or any combination of these movements.

This animation was created with a library of 3D development "OpenGL" which is a standard library for representing a three dimensional display of realistic scenes with some geometric transformations.

The embedded software overview flowchart is illustrated in Fig. 7. First, the communication protocols are reset, and sensors correctly set. Then, the sensor unit is configured (range, sampling frequency, and internal filter cut off frequency selected

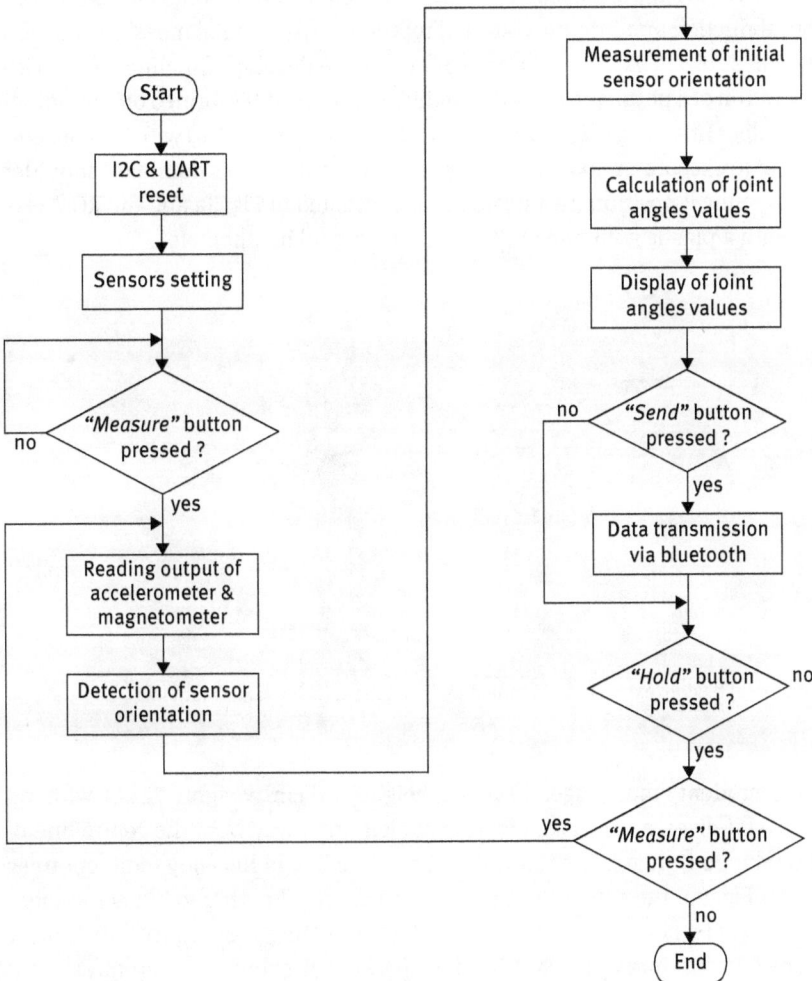

Fig. 7. Embedded software overview flowchart.

to be adapted to our application) before being attached to the segment associated with the investigated movement. The next step consists in testing if the Measure button is pressed or not before launching the computation of flexion/extension, adduction/abduction and int./ext. rotation angles according to the corresponding equations presented in section 2.1. As the joint angles are obtained, the user can send data to GUI via a wireless Bluetooth transmission by pressing Send button, and/or read the results on the display screen.

3.3 Preliminary tests

Preliminary tests have been performed on the shoulder joint as it has three degrees of freedom DOF. Three simple movements were considered: shoulder flexion/extension, adduction/abduction and internal/external rotation. Experimental measurements on three different subjects have been carried out using the developed inclinometer. With the collaboration of a physician in sports medicine, a test was performed on a handball athlete volunteer (male, age: 23 years, height: 170 cm, weight: 80 kg) with his consent. He was asked to execute a maximal active flexion/extension of his dominant shoulder from the anatomical position (rest position) as presented in Fig. 8a and the ROM was measured with a plastic goniometer then with the digital inclinometer.

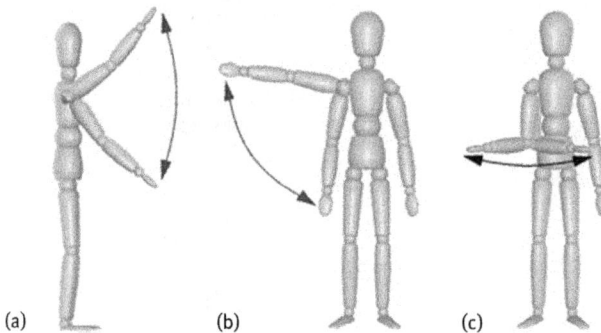

Fig. 8. Shoulder (a) flexion/extension; (b)adduction/abduction; (c) int./ext. rotation.

A volunteer student (male, age: 23 years, height: 185 cm, weight: 72 kg) with no musculoskeletal disease was asked to take his left arm away from the centre line of his body (abduction) then bring it towards the centre line of his body (adduction) as illustrated in Fig. 8b, the sensor unit placed on his arm. Another volunteer student (male, age: 25 years, height: 175 cm, weight: 60 kg) in sitting position with the sensor unit attached to his forearm, was asked to perform a shoulder internal/external rotation as shown in Fig. 8c.

4 Results and discussion

4.1 Calibration results

Static calibration graphs of pitch and roll angles measured by the accelerometer are represented in Fig. 9a and Fig. 9b, respectively. One can notice that experimental data exhibit linear behavior in both cases. Experimental data corresponding to pitch angles in Fig. 9a has been fitted by a straight red line (slope equal to 0.99) and those corresponding to roll angles in Fig. 9b were also fitted by a straight red line (slope equal to 1.01). The correlation coefficients were found to be equal to 0.9997 in both cases. The slope values can be approximated to 1 in the two cases.

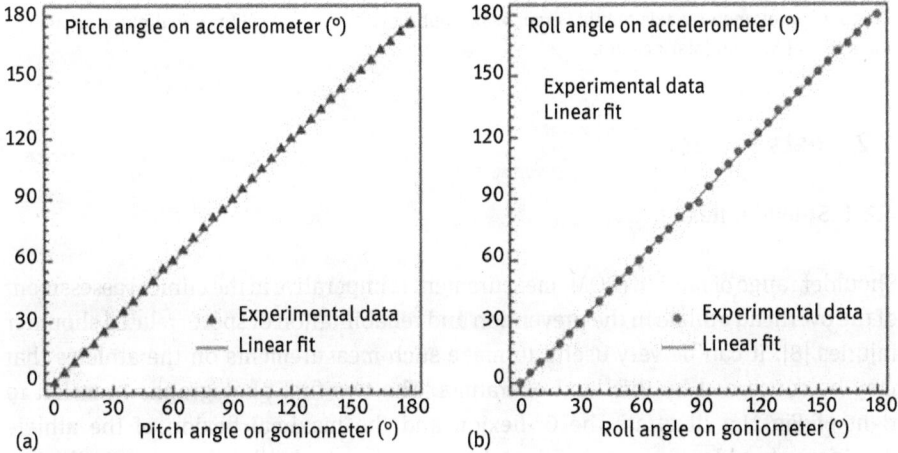

Fig. 9. Static calibration curve of (a) pitch angle, (b) roll angle of the accelerometer.

The magnetometer calibration curveis illustrated in Fig. 10. The circular pattern of light-gray graph representing $Y_h = f(X_h)$ indicates that only hard-iron interference magnetic field was present. The black circle corresponds to the calibrated magnetometer, indicating that hard-iron effect has been compensated. It has been centered at origin (0,0) after eliminating the offsets associated with hard-iron interference magnetic field. The offset values on the two axes X and Y have been removed by software. Soft-iron effect was eliminated by placing the sensor in a separate unit (sensor unit) far away from the electronic components inside the main unit.

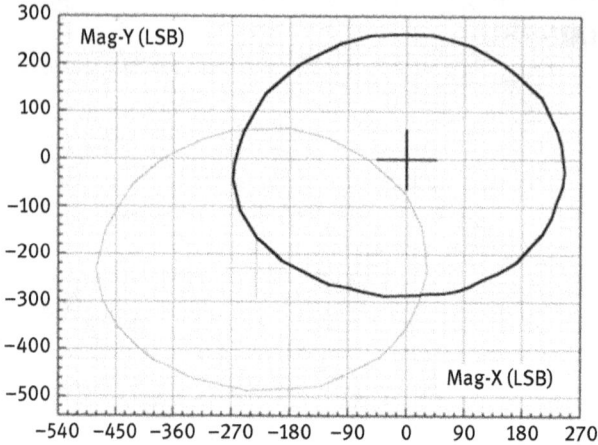

Fig. 10. Magnetometer *y*-axis output versus *x*-axis output (a.u.) :before calibration (light-gray circle) and after calibration (black circle).

4.2 Tests results

4.2.1 Shoulder flexion

Shoulder range of motion (ROM) measurement is imperative in the clinical assessment of the overhead athlete in the prevention and rehabilitation of sports related shoulder injuries [8]. It can be very useful to make such measurements on the athletes that may be concerned by this kind of trauma. The two first photographs from left to right of Fig. 11.a illustrate the 0°-flexion and the maximal flexion of the athlete dominant shoulder measured with a plastic goniometer by the physician. In the two last photographs, the same procedure is shown with the shoulder flexion measured usingthe digital inclinometer (sensor unit attached to a plastic ruler).

The graph presented in Fig. 11.b shows the joint angle versus time when the athlete in Fig. 11.a performed the flexion-extension movements in the sagittal plane.

The rest position corresponding to anatomical position defined zero. Flexion was defined as positive and extension as negative angles. The maximal flexion obtained from the digital inclinometer was equal to 160° (variation on the peak was considered as an artefact) and the maximal extension flexion was equal to 80°. The corresponding angles obtained from the plastic goniometer were 163° and 78°. One can notice that both devices give approximately the same range values. The slight difference between the results from goniometer and the inclinometer are probably due to a not precise positioning of the device.

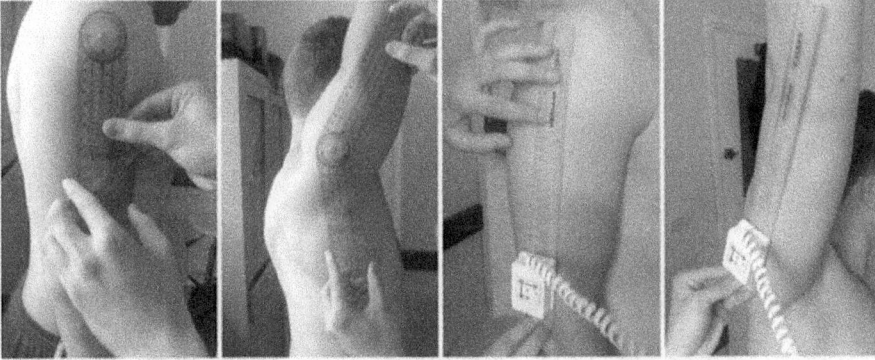

(a)

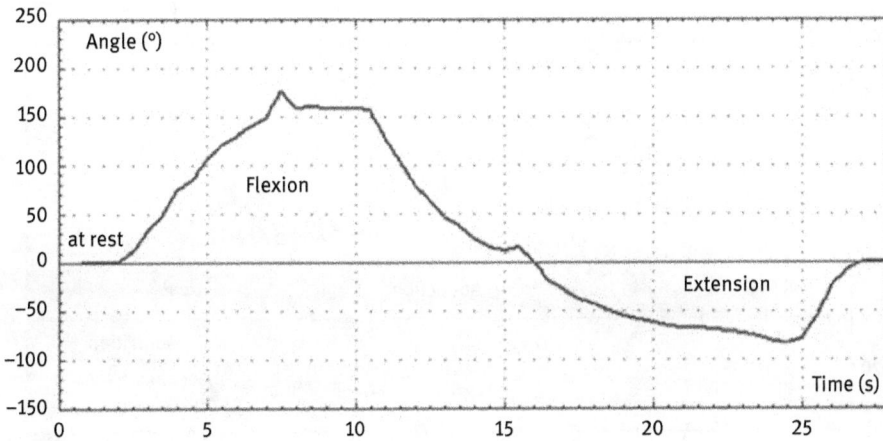

(b)

Fig. 11. (a): From left to right: Rest position and maximal flexion measured with plastic goniometerand digital inclinometer. (b): Dominant shoulder flexion-extension versus time performed by the athlete.

4.2.2 Shoulder adduction-abduction

From the left to right, the first photograph of Fig. 12.a corresponds to the arm neutral position, the second one to the 90°-abduction of the student left shoulder and the last one to the shoulder adduction that have taken place in the frontal plane.

The graph presented in Fig. 12.b presents the joint angle evolution versus time when the subject in Fig. 12.a performed the abduction-adduction movements in the frontal plane. The rest position corresponding to the arm neutral position defined zero.

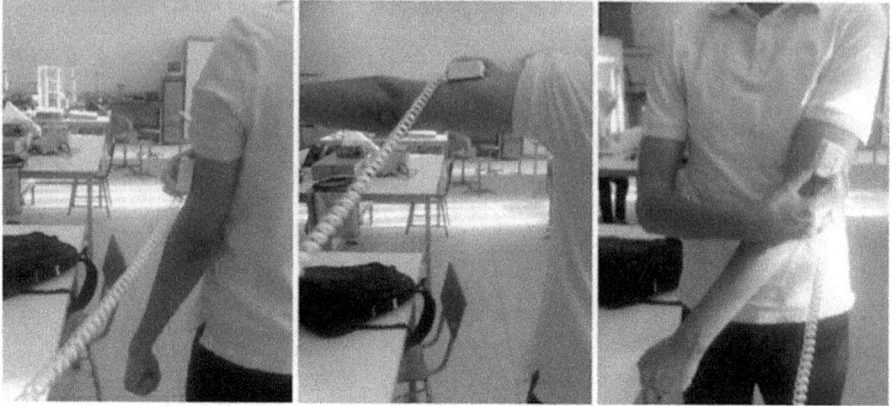

(a)

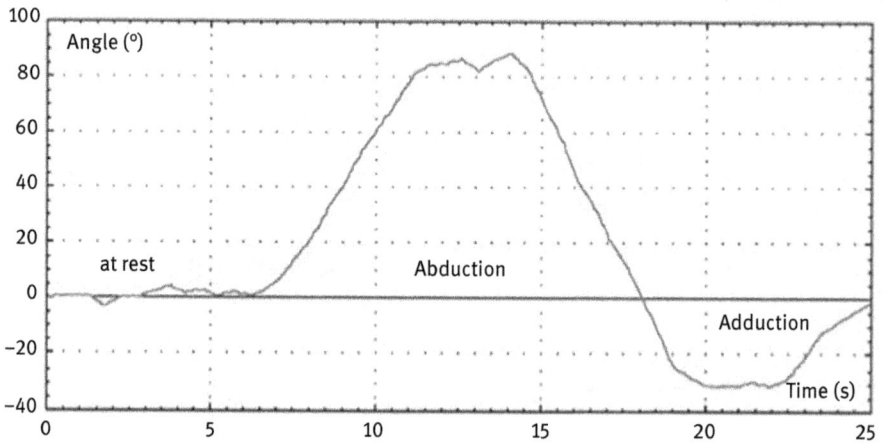

(b)

Fig. 12. (a): From left to right: Rest position, 90°-abduction and adduction of the shoulder.
(b): Shoulder abduction/adduction versus time.

Abduction was defined as positive and adduction as negative angles. The measured abduction angle value with the inclinometer was found to be equal 90° (variation on the peak was considered as an artefact) and the measured adduction angle value was found to be equal to 30°.

4.2.3 Shoulder internal-external rotation

From the left to right, the first photograph of Fig. 13.a corresponds to the arm neutral position with 0°-flexion of the elbow, the second one to the left shoulder internal rotation, and the last one to the shoulder external rotation.

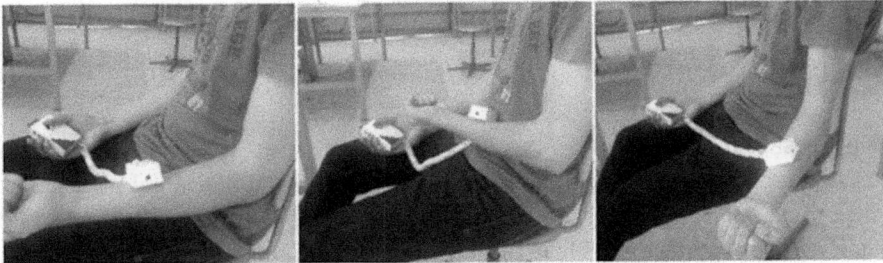

(a)

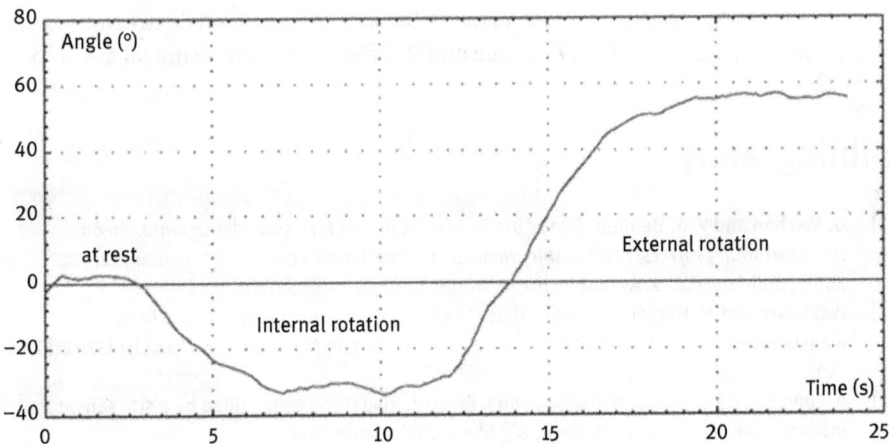

(b)

Fig. 13. (a): From left to right: Rest position, internal and external rotation of the shoulder. (b): Shoulder internal/external rotation versus time.

Shoulder internal/external rotation angle performed by the subject in Fig. 13.a versus time graph is depicted in Fig. 13.b. Contrary to the two previous movements that wereachieved in vertical planes, the internal/external rotation movement was performed in the horizontal transverse plane. These results cannot be obtained with a classical inclinometer because the subject is sitting, and not in supine position. The rest position corresponding to the arm neutral position defined zero. External rotation was defined as positive and internal rotation as negative angles. The measured internal rotation angle value with the inclinometer was found equal to 35° while the measured external rotation angle value was found equal to 56°.

According to the sensor circuit manufacturer, heading accuracy is below 2°, and pitch and roll accuracy are below 1° when tilt is below 50° [21]. In this study, the inclinometer accuracy can be considered as equal to 1° for measurements in the vertical plane and 2° in the horizontal plane.

5 Conclusion

A low cost digital inclinometer has been developed in our laboratory in order to measure range of motion simultaneously in the vertical and horizontal planes. The range of motion can be displayed on the inclinometer screen or transmitted to PC where a GUI can provide a real time graphical display of the measured angles as well as a 3D-animation corresponding to the movement performed by the subject in 3D-space.

Acknowledgment: The authors wish to thank Dr. S. Bencheikh from Centre National de Médecine du Sport (Algiers), T. Mahiouz and R. Achour for their technical assistance.

Bibliography

[1]   G. Machan and V. G. Bennett. *Use of inclinometers for geotechnical instrumentation on transportation projects: state of the practice.* Transportation Research E- Circular :92, 2008.[Online]. E-C129. Available: http://worldcat.org/issn/00978515

[2]   P. O'Leary and M. Harker. A framework for the evaluation of inclinometer data in the measurement of structures. *IEEE Trans. Instrumentation and Measurement*, 61(5):1237–1251, 2012.

[3]   A. Lambert, P. Furgale, T. D. Barfoot and J. Enright. Visual odometry aided by a sun sensor and inclinometer. *IEEE Aerospace Conf.*, Big Sky – USA, March 2011.

[4]   F. Baronti, G. Fantechi, R. Roncella and R. Saletti. A new and accurate system for measuring cruising yacht freeboards with magnetostrictive sensors. *IEEE Trans. Instrumentation and Measurement*, 60(5):1811–1819, 2011.

[5]   R. Thomas, G. Sen Gupta and K. Mercer. Sensors for an omni-directional mobile platform.In *IEEE Instrumentation and Measurement Technology Conference* (I2MTC), Binjiang – China, May 2011.

[6]   H.J. Luinge, P.H. Veltink and C.T.M. Batenc. Ambulatory measurement of arm orientation. *Journal of Biomechanics*, 40(1):78–85, 2007.

[7]   J. S. Scibek and C. R. Carcia. Assessment of scapulohumeral rhythm for scapular plane shoulder elevation using a modified digital inclinometer. *World Journal of Orthopedics*, 3(6):87–94, 2012.

[8]   A. Cools, L. De Wilde, A. van Tongel and D. Cambier. Measuring shoulder external and internal rotation strength with a hand-held dynamometer, and range of motion using a goniometer and a digital inclinometer: comprehensive intra- and inter rater reliability study of several testing protocols. *Bristish Journal of Sports Medicine*, 48(7):580–581, 2014.

[9]   N. J. MacIntyre, A. L. Lorbergs and J. D. Adachi. Inclinometer-based measures of standing posture in older adults with low bone mass are reliable and associated with self-reported, but not performance-based, physical function. *Osteoporosis Int.*, 25(2):721–728, 2014.

[10] G. C. David.Ergonomic methods for assessing exposure to riskfactors for work-related musculoskeletal disorders. *Occupational Medicine*, 55(3):190–199, 2005.

[11] P. Cheng and B.Oelmann. Joint-angle measurement using accelerometers and gyroscopes - a survey. *IEEE Trans. Instrumentation and Measurement*, 59(2):404–414, 2010.

[12] A. Brennan, J. Zhang, K. Deluzio and Q. Li. Quantification of inertial sensor-based 3D joint angle measurement accuracy using aninstrumented gimbal. *Gait & Posture*, 34(3):320–323, 2011.

[13] M.Yazdifar, M. R.Yazdifar, J.Mahmud, I.Esat and M.Chizari. Evaluating the hip range of motion using the goniometer and video tracking methods. *Procedia Engineering*, 68:77–82, 2013.

[14] T. Rudolfsson, M. Björklund and M. Djupsjöbacka. Range of motion in the upper and lower cervical spine in people with chronic neck pain. *Manual therapy*, 17(1):53–59, 2012.

[15] E. Seminati, A. Marzaric, O. Vacondiod and A.E. Minetti. Shoulder 3D range of motion and humerus rotation in two volleyball spike techniques: injury prevention and performance. *Sports Biomechanics*, 14(2):216–231, 2015.

[16] G. Myklebust, L. Hasslan, R. Bahr and K. Steffen. High prevalence of shoulder pain among elite Norwegian female handball players. *Scandinavian Journal of Medicine & Science in Sports*, 23(3):288–294, 2013.

[17] G.P.L. Almeida, P.F. Silveira, N.P. Rosseto, G. Barbosa, B. Ejnisman and M. Cohen. Glenohumeral range of motion in handball players with and without throwing-related shoulder pain. *Journal of Shoulder and Elbow Surgery*, 22(5):602–607, 2013.

[18] M.J. Kolber and W.J. Hanney. The reliability and concurrent validity of shoulder mobility measurements using a digital inclinometer and goniometer: A technical report. *Int. Journal of Sports Physical Therapy*, 7(3):306–313, 2012.

[19] D.C Hannah and J. S Scibek. Collecting shoulder kinematics with electromagnetic tracking systems and digital inclinometers: A review. *World Journal of Orthopedics*, 6(10):783–794, 2015.

[20] K. Tuck. Tilt Sensing Using Linear Accelerometers. *Freescale Semiconductor Application*, Note AN3461 Rev. 2, 06/2007.

[21] ST Microelectronics. *Using LSM303DLH for a tilt compensated electronic compass*. Application Note AN3192, 2010.

Biographies

Mahdi Zabat was born in Tebessa, Algeria, in 1991. He received the Bachelor of Engineering degree in Electrical Engineering in 2011, and the Master of Science degree in Electronic Instrumentation Engineering in 2013 from the University of Science and Technology Houari Boumediene (USTHB) in Algiers, Algeria. He is currently a Ph.D. student in Electronic Instrumentation at the laboratory of Instrumentation with the Faculty of Electronics and Computer Science, USTHB. His current research interests include the human joint angles measurements with different methods, inertial sensors and embedded systems.

Nazim Ouadahi was born in Algiers, Algeria, in 1988. He received the Bachelor of Engineering degree in Electrical Engineering in 2010 and the Master of Science degree in Electronic Instrumentation Engineering in 2012 from the University of Science and Technology Houari Boumediene USTHB at Algiers. Since then, he has been PhD student in Electronic Instrumentation at the laboratory of Instrumentation with the Faculty of Electronics and Computer Science, USTHB. He is currently working on different electronic and embedded devices to study biomechanical parameters and their influence in sport performance assessment and virtual reality systems as well as in rehabilitation applications.

Amina Ababou received the B.Sc. degree in Physics from the University of Science and Technology of Algiers, Algiers, Algeria, in 1982, the M.Sc. and the Ph.D. degrees in Physics from Claude Bernard Lyon I University, Lyon, France, in 1983 and 1986 respectively. She joined the University of Science and Technology Houari Boumediene (USTHB) where she is currently a Professor and a researcher with the Instrumentation Laboratory. She teaches courses in physical sensors and smart materials. Her research interests include movement analysis, sensors and ambulatory systems. She is committed to collaborations with physicians in sports medicine and rehabilitation at the Centre National de Médecine Sportive (CNMS) on projects related to the management of disables and development of equipment dedicated to assisted rehabilitation.

Noureddine Ababou received the B.Sc. degree in Physics in 1980 from the University of Science and Technology of Algiers, Algiers, Algeria, the M.Sc.and the Ph.D degree in Physics from Joseph Fourier University, Grenoble, France. In 1983, he joined the University of Science and Technology Houari Boumediene (USTHB), Algiers. He is currently Associate Professor atthe Faculty of Electronics and Computer Science at USTHB where he is teaching measurement devices, sensors and movement analysis at undergraduate and graduate levels and he is a researcher with the laboratory of Instrumentation. He has been visiting Professor from 2001 to 2011 at the INFS-STS sport technology institute, Algiers. His main research interests include sensors, rehabilitation devices, movement analysis, ambulatory systems, and protective equipments in sports.

Karim Ben Si Said was born in Algeria in 1990. He received the Bachelor of Engineering degree in Electrical Engineering in 2011 and the Master of Science degree in Electronic Instrumentation Engineering in 2013 from the University of Sciences and Technology Houari Boumediene USTHB, Algiers, Algeria. Currently, he is a Ph.D. student in Electronic Instrumentation at the laboratory of Instrumentation with the Faculty of Electronics and Computer Science, USTHB. His main areas of research interest are the instrumentation dedicated to inter-segmental coordination analysis in sport and rehabilitation, and the instrumentation dedicated to virtual reality.

Arezki Youyou was born in Algeria in 1988. He received the Bachelor of Engineering degree in Electrical Engineering in 2009, and the Master of Science degree in Electronic Instrumentation Engineering in 2011 from the University of Science and Technology Houari Boumediene (USTHB) in Algiers, Algeria. He is currently a PhD student in Electronic instrumentation at the laboratory of Instrumentation with the Faculty of Electronics and Computer Science, USTHB. His field of research focused on biomechanical instrumentation, and embedded systems that allow the monitoring and analysis of the human movement.

S. Ben Sassi and F. Najar

Accurate Reduced-Order Modeling of MEMS and NEMS Microactuators under Dynamic Electrostatic Loading and Large Strokes

Abstract: When modeling micro/nano electromechanical systems (MEMS/NEMS), taking into account of geometric nonlinear effects can be sometimes ambiguous. In this study we investigate the static and dynamic behavior of micro and nano beams based on MEMS and NEMS, when fully integrating nonlinear terms. We demonstrate that neglecting third-order nonlinearities in the equation of motion leads to accurate results and low computational costs in some cases. We start by deriving the governing equation of motion using both linear and nonlinear Euler Bernoulli beam theory with two possible configurations: cantilever and doubly-clamped beams. A Reduced Order Model (ROM) based on Differential Quadratic Method (DQM) is utilized to simulate the static response of the beam. Besides, we employ the Finite Difference Method (FDM) to discretize the orbits of motion and solve the resulting nonlinear algebraic dynamic equations for limit-cycle solutions. A comparison is then conducted between findings of the DQM-FDM decomposition and previous published results.

Keywords: MEMS, NEMS, geometric nonlinearities, DQM, FDM

1 Introduction

Recent years witnessed an explosion in the MEMS/NEMS market consumer. In fact, micro/nano devices attract more and more attention thanks to their great features which are principally small size, weight and low energy consumption. Also, the electro-mechanical coupling of MEMS is of particular interest since it is often exploited for sensing or actuation [1]. As a matter of fact, it has been shown that some physical nonlinear phenomena such as jump or pull-in appear on the system when this later is actuated using electrostatic forces at the micro/nano scale. Hence, accurate Reduced-Order Model (ROM) must be adopted in order to properly capture the different dynamic phenomena of MEMS/NEMS.

Mechanical nonlinearities in MEMS/NEMS devices can be caused either by material nonlinearity which is connected to the inelastic aspect of the material, or by geometric nonlinearity, related to large defection of structures. The latter type is principally generated when the displacement becomes very large compared to the

S. Ben Sassi and F. Najar: Applied Mechanics and Systems Research Laboratory, Tunisia Polytechnic School, La Marsa, University of Carthage, Tunisia. Email: fehmi.najar@ept.rnu.tn.

De Gruyter Oldenbourg, ASSD – Advances in Systems, Signals and Devices, Volume 6, 2018, pp. 21–34.
https://doi.org/10.1515/9783110448375-002

structure dimensions, resulting in a nonlinear relationship between the extension of the structure and the bending vibrations [2]. In the previous literature performed on modeling of MEMS and NEMS devices, a few number of studies had focused on the effect of third order nonlinearities such as nonlinear curvature and inertia. For instance, Chatarjee and Pohit [3] and Ouakad [4] in the micro scale, and Rasekh and Khadem [6] and Wang and Wang [5] in the nano scale, reported the influence of third order nonlinearities in the case of large displacements or when the gap length ratio is superior to 0.3. Collenz et al. [7] developed a new FEM method by considering very large displacements of a micro cantilever beam. Rajabi and Ramezani [8] demonstrated that geometric nonlinearity amplifies the natural frequency of a micro cantilever beam when increasing the beam thickness.

On the other hand, several studies were focused on model order reduction of MEMS/NEMS. Here is a brief overview of the main contributions. Younis et al. [9] presented a ROM based on Galerkin decomposition and a discretized Taylor series to handle electrostatic forces. The model was designed to study the static and dynamic response of an electrically actuated microbeam. The authors showed that solution convergence is achieved when employing at least three symmetric mode shapes. Samaali et al. [10] used the DQM to study the static and dynamic behaviors of a capacitive MEMS switch with double clamped micro-beams. They demonstrated that using only nine grid points in DQM leads to accurate results. Nayfeh et al. [11] studied the dynamic pull in of a doubly clamped microbeam in order to design MEMS resonators and micro switches. They used a ROM based on Galerkin approximation and the shooting method in order to simulate the microbeam frequency and force responses.

From the above literature review, one can remark that some ROMs are time consuming especially when all sources of nonlinearities are included in the model. On the other hand, neglecting third order nonlinearities may lead to erroneous results especially for large deflection of the beam. In this study we develop a ROM for both moderate and large displacements based on FDM-DQM discretization. Besides, we study the effects of including geometric and inertial nonlinearities in the micro and nano scales, and prove the accuracy and efficiency of the developed ROM for large displacements even when neglecting third order nonlinearities. We start by formulating the problem using both linear and nonlinear beam theory with two possible configuration types: cantilever beam and doubly-clamped beam. As a ROM, DQM is employed to simulate the static response of the micro and nano beams. Then, we apply a combination of FDM-DQM to calculate limit-cycle solutions for beam theories. A comparison with previous published studies is performed to validate the proposed ROM.

2 Problem formulation

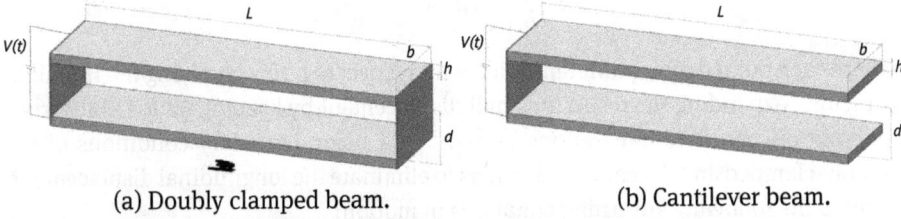

(a) Doubly clamped beam.   (b) Cantilever beam.

Fig. 1. Schematic of electrically actuated MEMS/NEMS devices.

We propose to study the static and dynamic behavior of micro and nano beams subjected to electrostatic voltage. The structures under consideration have two configurations: doubly-clamped (Clamped-Clamped CC) beam and cantilever (Clamped-Free CF) beam as shown in Fig. 1. The system is composed by an upper flexible electrode connected to a voltage source and fixed substrate. We use the nonlinear Euler-Bernoulli beam theory in order to derive the governing equation of motion for the two types beam configurations. The beam extension and rotation are given by [2]:

$$e = u'(x, t) + \frac{1}{2}w'(x, t)^2 - \frac{1}{2}u'(x, t)w'(x, t)^2 \tag{1}$$

$$\theta = w'(x, t) - u'(x, t)w'(x, t) \tag{2}$$

where $w(x, t)$ is the deflection of the upper beam at time t and at locations x and $u(x, t)$ represents the axial displacement. Here each prime (') denotes differentiation with respect to x. We express the variation of the virtual kinetic energy δK, the virtual elastic energy δU and virtual potential electrostatic energy δV_e as:

$$\delta K = \int_0^L \rho bh \left(\dot{u}\delta\dot{u} + \dot{w}\delta\dot{w}\right) + J(w' - u'w')\left((1 - u')\,\delta w' - w'\,\delta u'\right)dx \tag{3}$$

$$\delta U = \int_0^L EA \left(u' + \frac{1}{2}w' - \frac{1}{2}u'w'^2\right)\left(1 - \frac{1}{2}w'^2\right)\delta u' + EA\left(1 - u'\right)w'\delta w'$$

$$+ EI\left(w'' - u'w'' - u''w'\right)\left(\delta w''\left(1 - u'\right) - \delta u'w'' - \delta u''w' - u''\delta w'\right)$$

$$+ c\frac{\partial w}{\partial t}\,\delta w\,dx \tag{4}$$

$$\delta V_e = \int_0^L \frac{\varepsilon b (V_{dc} + v(t))^2}{(d-w)^2} \delta w \, dx \tag{5}$$

where each point denotes differentiation with respect to t. Now applying the Hamilton principle, expanding the result for small displacement by keeping up to cubic terms, carrying out multiple integrations by parts and using boundary conditions of the doubly-clamped and the cantilever beams to eliminate the longitudinal displacement, yield to the following governing equations of motion:

Equation of motion of the doubly-clamped beam:

$$\ddot{w} + \bar{c}\dot{w} + w^{iv} = \left(w'' + \alpha_{nl}\alpha_4 w^{iv} \right)\left(\bar{N} + \alpha_1 \int_0^L w'^2 \, dx \right)$$

$$-\alpha_3 \alpha_{nl}\left(w''^3 + 2w'w''w''' \right) + \alpha_2 \frac{(V_{dc} + v(t))^2}{(1-w)^2} \tag{6}$$

with the boundary conditions

$$w(0,t) = 0, \quad w(1,t) = 0, \quad w'(0,t) = 0 \quad \text{and} \quad w'(1,t) = 0 \tag{7}$$

Equation of motion of the cantilever beam:

$$\ddot{w} + \bar{c}\dot{w} + w^{iv} = -\alpha_1 \alpha_{nl}\left(w''^3 + 4w'w''w''' + w^{iv}w'^2 \right) + \frac{\alpha_2 (V_{dc} + v(t))^2}{(1-w)^2}$$

$$-\frac{\alpha_{nl}\alpha_1}{2}\left(w'' \int_1^x \int_0^x \left(w'^2 \right)^{\cdot\cdot} dx\,dx + w' \int_1^x \left(w'^2 \right)^{\cdot\cdot} dx \right) \tag{8}$$

with the boundary conditions

$$w(0,t) = 0, \quad w'(0,t) = 0, \quad w''(1,t) = 0 \quad \text{and} \quad w'''(1,t) = 0 \tag{9}$$

Where $\bar{c}$ represents the beam's damping related to the quality factor by $Q = \frac{\omega}{\bar{c}}$, ω is the non-dimensional natural frequency, $v(t) = V_{AC} \times \cos(\Omega t)$ is the applied voltage. α_{nl} is a variable that takes 0 when considering only the Von Karman nonlinearity in the Euler-Bernoulli theory and 1 when using a higher order nonlinear theory. For convenience, we use the following non-dimensional variables:

$$\tau^2 = \frac{\rho bh L^4}{EI}, \quad \bar{c} = \frac{c\tau}{\rho bh}, \quad \alpha_1 = 6\left(\frac{d}{h}\right)^2,$$

$$\alpha_2 = 6\frac{\varepsilon L^4}{Eh^3 d^3}, \quad \alpha_3 = \frac{d^2}{2L^3}, \quad \alpha_4 = \frac{h^2}{6L^2} \tag{10}$$

where ρ is the density, E is the modulus of elasticity, b, h, $A = bh$ and $I = bh^3/12$ are beam's width, thickness, cross-section area and second moment of area, respectively. d is the initial gap distance between both beams and $\bar{N} = \frac{NL^2}{EI}$ represents the non-dimensional axial force representing the residual stresses. Equation 6 contains three types of nonlinearities: the electrostatic force, the nonlinear curvature and the midplane stretching. While Equation 8 contains nonlinearities due to the electrostatic force, curvature and inertia. In the present work, we aim at studying how can nonlinear curvature and inertia affect the static and dynamic response of beams at both micro and nano scales. For that, we set a constant a_{nl} that can takes $\{0, 1\}$ to account or not for these nonlinear terms.

3 Reduced Order Modeling

To simulate the static and dynamic response of the microbeam, we use a ROM based on a combined discretization in time and space using DQM and FDM methods [17]. The DQ method is used alone for the static response. This method transforms a derivative into a weighted linear sum of n values of the function to be derived at the Chebyshev-Gauss-Lobatto grid points $\zeta_i = \frac{1}{2}\left(1 - \cos\frac{(i-1)}{(n-1)}\pi\right)$ for $1 \leq i \leq n$. Hence, the expression of the derivative becomes:

$$\left[\frac{\partial w^r(x)}{\partial x^r}\right]_{x=x_i} = \sum_{j=1}^{n} A_{ij}^{(r)} w_j \tag{11}$$

where the weighting coefficient are determined using the Lagrange interpolation polynomials. They are given by:

$$A_{ij}^{(1)} = \left(\frac{\displaystyle\prod_{v=1;v\neq i}^{n}(\zeta_i - \zeta_v)}{(\zeta_i - \zeta_j)\displaystyle\prod_{v=1;v\neq j}^{n}(\zeta_j - \zeta_v)}\right) \quad i,j = 1,2,...,n, \;\; j\neq i \tag{12}$$

$$A_{ij}^{(r)} = r\left(A_{ii}^{r-1}A_{ij}^{1} - \frac{A_{ij}^{r-1}}{(\zeta_i - \zeta_j)}\right) \quad i,j = 1,2,...,n, \;\; j\neq i \tag{13}$$

$$A_{ii}^{(r)} = -\sum_{v=1;v\neq i}^{n} A_{iv}^{(r)} \quad i = 1,2,...,n \tag{14}$$

Thus, discretizing the governing equations of motion will yield:
Equation of motion of the clamped beam: ($i = 3, ..., n-2$)

$$\sum_{j=1}^{n} A_{ij}^{(4)} w_j + \alpha_3 \alpha_{nl} \left\{ \left(\sum_{j=1}^{n} A_{ij}^{(2)} w_j \right)^3 + 2 \sum_{j=1}^{n} \sum_{k=1}^{n} \sum_{l=1}^{n} A_{ij}^{(1)} A_{ik}^{(2)} A_{il}^{(3)} w_j w_k w_l \right\}$$

$$= \left(\sum_{j=1}^{n} A_{ij}^{(2)} w_j + \alpha_{nl} \alpha_4 \sum_{j=1}^{n} A_{ij}^{(4)} w_j \right) \left(N + \alpha_1 \sum_{k=1}^{n} C_k \left(\sum_{l=1}^{n} A_{kl}^{(1)} w_l \right)^2 \right)$$

$$- \ddot{w}_i - c\dot{w}_i + \alpha_2 \frac{(V_{dc} + v(t))^2}{(1 - w_i)^2} \tag{15}$$

$$w_1 = 0, \quad w_n = 0, \quad \sum_{j=1}^{n} A_{1j}^{(1)} w_j = 0 \quad \text{and} \quad \sum_{j=1}^{n} A_{nj}^{(1)} w_j = 0 \tag{16}$$

Equation of motion of the cantilever beam: $(i = 3, \ldots, n - 2)$

$$\ddot{w}_i + \sum_{j=1}^{n} A_{ij}^{(4)} w_j + \alpha_3 \alpha_{nl} \left\{ \left(\sum_{j=1}^{n} A_{ij}^{(2)} w_j \right)^3 + 4 \sum_{j=1}^{n} \sum_{k=1}^{n} \sum_{l=1}^{n} A_{ij}^{(1)} A_{ik}^{(2)} A_{il}^{(3)} w_j w_k w_l \right.$$

$$+ \sum_{j=1}^{n} A_{ij}^{(4)} w_j \left(\sum_{k=1}^{n} A_{ik}^{(1)} w_k \right)^2 \right\} + c\dot{w}_i = \alpha_3 \alpha_{nl} \left\{ \sum_{j=1}^{n} \sum_{k=j}^{n} \sum_{l=1}^{n} D_{jk} E_{kl} A_{ij}^{(2)} w_j \right.$$

$$\left(\sum_{q=1}^{n} \sum_{p=1}^{n} A_{lq}^{(1)} A_{lp}^{(1)} w_q \ddot{w}_p + \left(\sum_{r=1}^{n} A_{lr}^{(1)} \dot{w}_r \right)^2 \right) + \sum_{j=1}^{n} \sum_{k=j}^{n} E_{jk} A_{ij}^{(1)} w_j$$

$$\left(\sum_{q=1}^{n} \sum_{p=1}^{n} A_{kq}^{(1)} A_{kp}^{(1)} w_q \ddot{w}_p + \left(\sum_{r=1}^{n} A_{kr}^{(1)} \dot{w}_r \right)^2 \right) \right\} + \alpha_2 \frac{(V_{dc} + v(t))^2}{(1 - w_i)^2} \tag{17}$$

$$w_1 = 0, \quad \sum_{j=1}^{n} A_{1j}^{(1)} w_j = 0, \quad \sum_{j=1}^{n} A_{nj}^{(2)} w_j = 0 \quad \text{and} \quad \sum_{j=1}^{n} A_{nj}^{(3)} w_j = 0 \tag{18}$$

It is worth to mention that integral terms are evaluated using Newton Cotes formula. Thus, the weighted coefficients C_j, D_{ij} and E_{ij} are derived from the Lagrange basis polynomials. That is:

$$C_i = \int_0^1 \prod_{r=1; r \neq i}^{n} \frac{x - \xi_r}{\xi_i - \xi_r} dx \quad i = 1, \ldots, n \tag{19}$$

$$D_{ij} = \int_{\xi_i}^1 \prod_{r=i; r \neq j}^{n} \frac{x - \xi_r}{\xi_j - \xi_r} dx \quad i, j = 1, \ldots, n \tag{20}$$

$$E_{ij} = \int_{0}^{\xi_k} \prod_{r=1;r\neq j}^{k} \frac{x-\xi_r}{\xi_j-\xi_r} dx \quad i,j,k = 1,...,n \tag{21}$$

To discretize the time, we use the FDM which basic principle is to discretize one orbit of period $T = \frac{2\pi}{\Omega}$ using $(m+1)$ points. Assuming a nonautonomous system, the period T of the solution is normalized and in order to insure the periodicity of the solution we enforce the condition: $w_{1j} = w_{nj}$. Using a central difference scheme, first and second derivative of the equation of motion can be written as:

$$w'_j(t) = \frac{w_{k+1j} - w_{k-1j}}{2h} \tag{22}$$

$$w''_j(t) = \frac{w_{k+1j} - 2w_{kj} + w_{k-1j}}{h^2} \tag{23}$$

where $h = \frac{T}{m}$ is the step size of the iterative scheme. Injecting Equations 22 and 23 in the equations of motion 15 and 17 will yield to the full discretized ROM.

4 Static response of the micro and nano beams under DC voltage

To obtain the static equation, we set time derivative terms equal to zero then we use DQM to solve the set of algebraic equations employing Newton-Raphson method. The residual stress N due to fabrication process is given by $156.165 \, \mu N$ for the double clamped micro beam and $1.35 \, \mu N$ for the double clamped nano beam.

Tab. 1. Micro and Nano beams parameters.

Beam	L (μm)	b (μm)	h (μm)	d (μm)	E (GPa)	ρ (kg/m³)
Micro C-C	510	100	1.5	1.18	166	2332
Micro C-F	210	100	1.5	1.18	166	2332
Nano C-C	15	0.15	0.1	0.3	226	3158
Nano C-F	0.2	0.03	3.510^{-3}	0.03	166	2332

Figure 2 and 3 show the voltage versus maximum deflection diagram of the doubly-clamped and cantilever beams in the micro and nano scales, respectively. From Figs 2a and 3a we remark a perfect agreement between curves when $\alpha_n = 0$ (line) and $\alpha_n = 1$ (points). Moreover, we notice a negligible effect of third order nonlinearities in the double clamped and cantilever nano beams through Fig. 2a and 3a. Thus, it can be concluded that nonlinearities due curvature and inertia do not affect the static response of a cantilever and double clamped beams in the micro and nano scales. For a low computational time and without affecting the accuracy of the obtained results, we deduce that one can neglect third order nonlinearities in the static analysis of nano and micro beams.

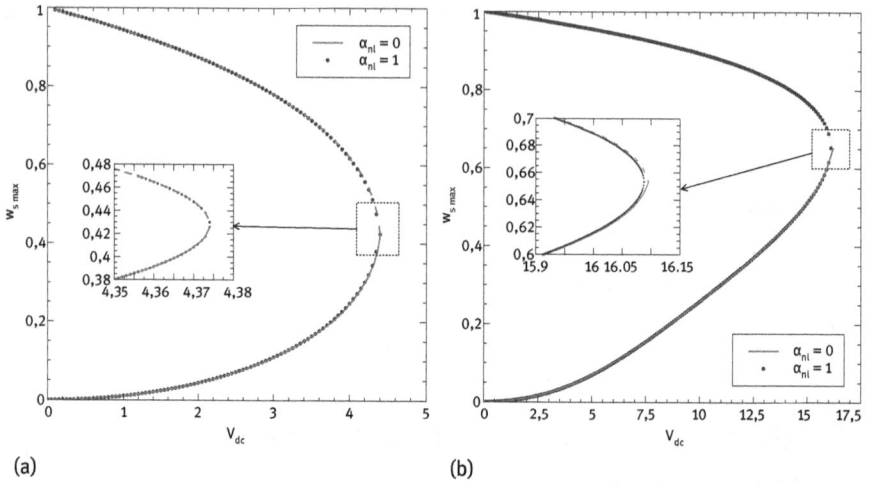

(a)   (b)

Fig. 2. Static analysis of the doubly clamped micro beams. (a) Micro beam. (b) Nano beam.

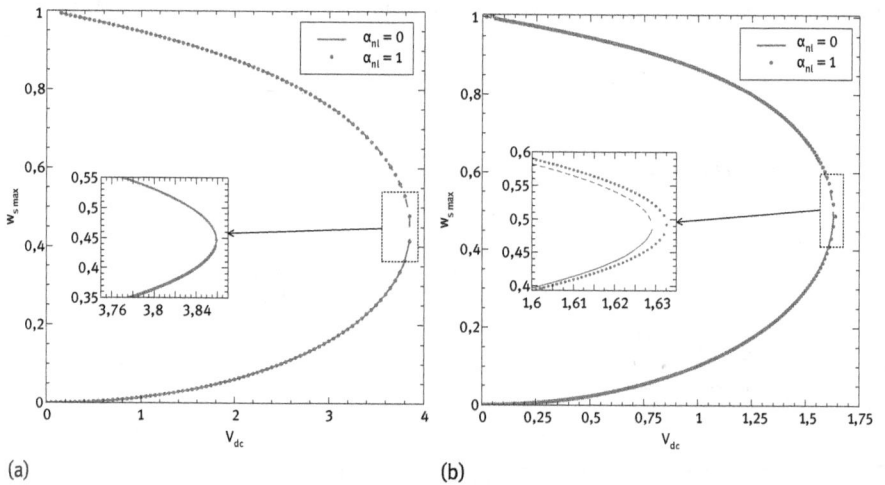

(a)   (b)

Fig. 3. Static analysis of the cantilever micro beams. (a) Micro beam. (b) Nano beam.

5 Dynamic response of the electrically actuated micro and nano beams

In order to accurately investigate the nonlinear behavior of the electrically actuated micro and nano beams, we simulate the dynamic response when the system is subjected to DC and AC voltage forcing. The system is excited nearby the fundamental frequency. We use the ROM described in section 3 then we solve the set of ODEs using Newton-Raphson method. Besides, we capture stable orbits using long time integration (LTI) using a 4th order Runge-Kutta technique.

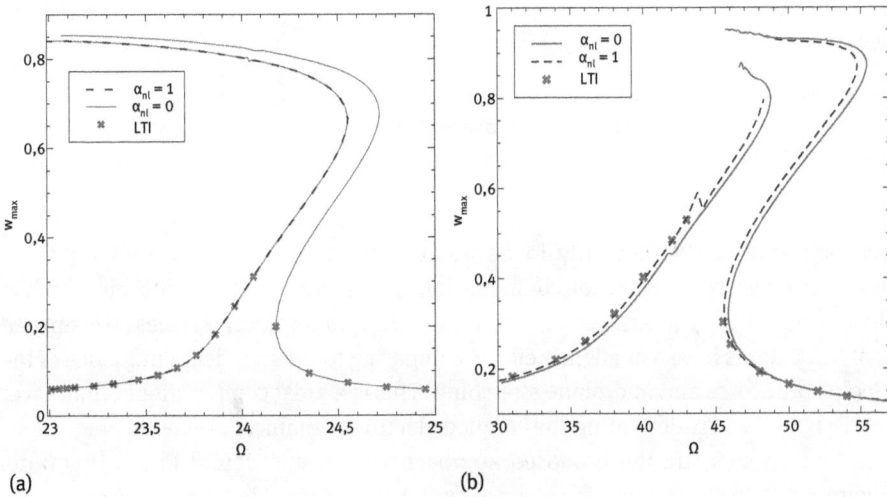

(a)   (b)

Fig. 4. Dynamic analysis of doubly clamped beams. (a) Frequency response of micro beam actuated by $V_{dc} = 2\,V$ and $V_{ac} = 0.1\,V$. (b) Frequency response of nano beam actuated by $V_{dc} = 10\,V$ and $V_{ac} = 2.5\,V$.

Figure 4 shows the frequency response curve of the doubly fixed beams in the micro scale in Fig. 4a and the nano scale in Fig. 4b. We remark that third order nonlinearities do not alter the dynamic response of the double clamped micro beam, while, in the case of nano beam we remark a slight deviation between curves for moderate and large displacements. Further, good agreement is observed between result given by LTI and FDM method. We recall that the global dynamic behavior of the double clamped beams is dominated by a softening type due to the high applied voltage. The local hardening behavior is due to the effect of mid-plane stretching.

Same remarks can be deduced for the cantilever beams cases. In fact, we notice through Figs 5a and 5b that both dashed ($\alpha_{nl} = 1$) and continuous curves ($\alpha_{nl} = 0$) are in

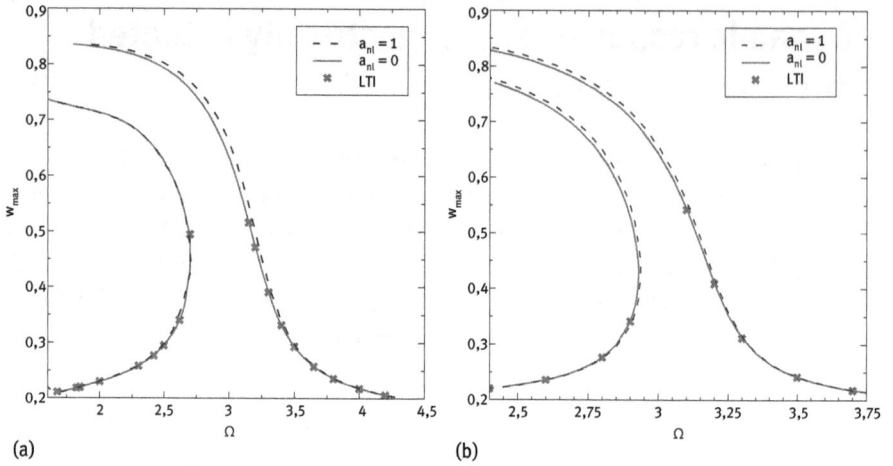

Fig. 5. Dynamic analysis of cantilevered beams. (a) Frequency response of micro beam actuated by $V_{dc} = 3\,V$ and $V_{ac} = 0.8\,V$. (b) Frequency response of nano beam actuated by $V_{dc} = 1.25\,V$ and $V_{ac} = 0.05\,V$.

good agreements. Consequently, for both static and dynamic responses of the system, and using two types of beam configuration in the micro and nano scale, we had demonstrated the uselessness of considering third-order nonlinearities. We showed that these terms have a negligible effects comparing to the significant influence of the electrostatic force and mid-plane stretching. This is a great computational time saver for designers when simulating the coupled electro-mechanical problem.

We also compare the proposed approach with results found in the literature. Figure 6a shows a frequency response function of a double clamped micro beam described in [11] where the author used a combination between 3 modes Galerkin projection and shooting method as a ROM. A perfect agreement is shown by superimposing results even for large displacement. The adopted ROM is also validated in the nano scale through comparison with frequency response of Ouakad et al. [12] where a Carbone Nanotube (CNT) is actuated by an electrostatic force given by the following expression:

$$F_{elec} = \frac{\pi \varepsilon_0 (V_{dc} + V_{ac} \cos(\Omega t))}{\sqrt{(d-w)(d-w+2R)} \left(\cosh^{-1}\left(1 + \frac{d-w}{R}\right)^2 \right)} \tag{24}$$

where R represents the radius of the CNT. The author used the Galerkin decomposition and the shooting method to solve the equation of motion. Figure 6b exhibits a good agreement between the two ROM for both low and moderately large displacement. We conclude the accuracy of the adopted ROM to predict the dynamic response of micro and nano beams based MEMS and NEMS.

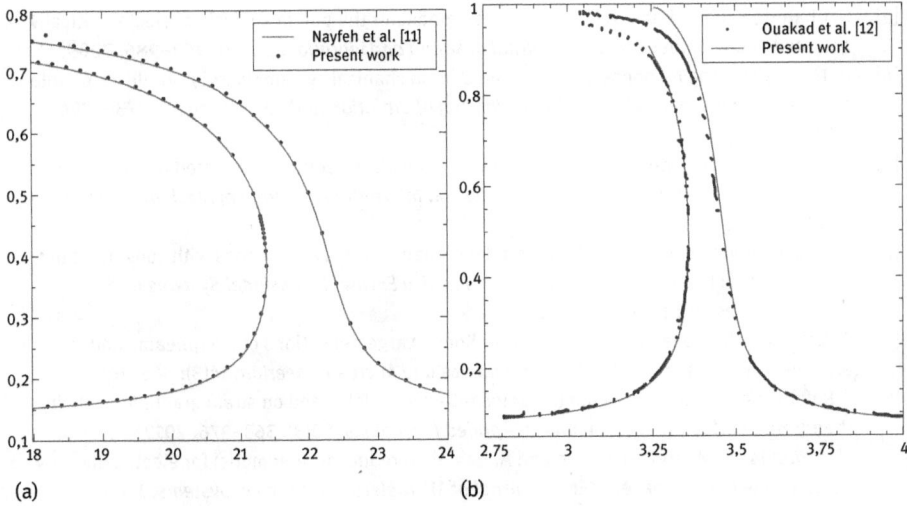

Fig. 6. Comparison of the adopted ROMs results and previous work. (a) Frequency response of doubly clamped micro beam in. (b) Frequency response of doubly clamped CNT beam in.

6 Conclusion

In this paper, we had constructed different modeling approaches for MEMS and NEMS devices in double clamped and cantilever type beam configurations. We had used linear and nonlinear beam theories to derive the governing equation of motion. We had simulated the static and dynamic response of the double clamped and cantilever configurations. Solving the obtained governing equations using DQM discretization for the static problem and DQM-FDM for the dynamic problem, we had demonstrated that both linear and nonlinear beam theories lead to the same results. We had concluded that neglecting third order nonlinear terms lead to a great computational time saver. Finding were also verified using numerical long time integration. Further, we had used the DQM-FDM with previously published MEMS and NEMS models. We had demonstrated that this technique is accurate and robust to predict the nonlinear dynamic behavior of MEMS and NEMS devices.

Bibliography

[1]   Mohammad I. Younis. *MEMS Linear and Nonlinear Statics and Dynamics: Linear and Nonlinear Statics and Dynamics.* Springer Science & Business Media, 2011.
[2]   Ali H. Nayfeh and Frank P. Pai. *Linear and nonlinear structural mechanics.* John Wiley & Sons, 2008.

[3]   S. Chaterjee and G. Pohit. A large deflection model for the pull-in analysis of electrostatically actuated microcantilever beams. *Journal of sound and vibration*, 322(4): 969–986, 2009.

[4]   H. M. Ouakad. The response of a micro-electro-mechanical system (MEMS) cantilever-paddle gas sensor to mechanical shock loads. *Journal of Vibration and Control*, 21(14): 273–2754, 2015.

[5]   M. Rasekh and S. E. Khadem. Pull-in analysis of an electrostatically actuated nano-cantilever beam with nonlinearity in curvature and inertia. *International Journal of Mechanical Sciences*, 53(2): 108–115, 2011.

[6]   K. F. Wang and B. L. Wang. A general model for nano-cantilever switches with consideration of surface effects and nonlinear curvature. *Physica E: Low-dimensional Systems and Nanostructures*, 66: 197–208, 2015.

[7]   A. Collenz, F. De Bona, A. Gugliotta and A. Somà. Large deflections of microbeams under electrostatic loads. *Journal of Micromechanics and Microengineering*, 14(3): 365, 2003.

[8]   F. Rajabi and S. Ramezani. A nonlinear microbeam model based on strain gradient elasticity theory with surface energy. *Archive of Applied Mechanics*, 82(3): 363–376, 2012.

[9]   M. I. Younis , E. M. Abdel-Rahman and A. Nayfeh. A reduced-order model for electrically actuated microbeam-based MEMS. *Journal of Microelectromechanical Systems*, 12(5): 672–680, 2003.

[10] H. Samaali, F. Najar and S. Choura. Dynamic study of a capacitive mems switch with double clamped-clamped microbeams. *Shock and Vibration*, 2014, 2014.

[11] A. H. Nayfeh, M. I. Younis and E. M. Abdel-Rahman. Dynamic pull-in phenomenon in MEMS resonators. *Nonlinear Dynamics*, 48(1–2): 153–163, 2007.

[12] H. M. Ouakad and M. I. Younis. Nonlinear dynamics of electrically actuated carbon nanotube resonators. *Journal of Computational and Nonlinear Dynamics*, 5(1): 011009, 2010.

[13] J. H Kuang and C. J. Chen. Dynamic characteristics of shaped micro-actuators solved using the differential quadrature method. *Journal of Micromechanics and Microengineering*, 14(4): 647, 2004.

[14] Najib Kacem. *Nonlinear dynamics of M&NEMS resonant sensors: design strategies for performance enhancement.* PhD thesis, Insa Lyon, Grenoble, 2010

[15] F. Najar, S. Choura, E. M. Abdel-Rahman, S. El-Borgi and A. H. Nayfeh. Dynamic Analysis of Variable-geometry Electrostatic Microactuators. *Journal of Micromechanics and Microengineering*, 16(11): 2449, 2006.

[16] F. Najar, A. H. Nayfeh, E. M. Abdel-Rahman, S. Choura and S. El-Borgi. Nonlinear Analysis of MEMS Electrostatic Microactuators: Primary and Secondary Resonances of the First Mode. *Journal of Vibration and Control*, 16(9): 1321–1349,2010.

[17] F. Najar, A. H. Nayfeh, E. M. Abdel-Rahman, S. Choura and S. El-Borgi. Dynamics and Global Stability of Beam-Based Electrostatic Microactuators. *Journal of Vibration and Control*, 16(5): 721–748, 2010.

Biographies

Sarah BEN SASSI received an engineering diploma in mechanical engineering from Tunisia Polytechnic School, University of Carthage in 2013. Since that year, she had enrolled as a PhD candidate in applied mechanics in the Tunisia Polytechnic School.She is interested on modeling and numerical simulation of MEMS and NEMS devices under different multi-physical coupled fields and on experimental characterization of various micro and nano structures.

Fehmi NAJAR Ph.D., is Associate Professor of Mechanical Engineering at the Tunisia Polytechnic School, University of Carthage. He received an Engineering Diploma in Mechanical Engineering from the National Engineering School of Tunis in 1997, and an M.S. in Structural Dynamics from the Ecole Centrale de Paris in 1998. After graduation, he worked as Technical Manager in the Graphic Art industry for a period of 6 years. He obtained his PhD in Mechanical Engineering in 2008 from the National Engineering School of Tunis, University of Tunis El Manar in collaboration with the Virginia Polytechnic School and State University (Virginia Tech) in USA. His research interests include MEMS and NEMS, structural dynamics, nonlinear dynamics, smart materials, energy harvesting, and multi-body dynamics.

M. Hadj Said, F. Tounsi, L. Rufer, B. Mezghani and M. Masmoudi

Dynamic Performance of a Narrow Frequency Band Acoustic Microsensor

Abstract: In this paper, we present a dynamic modeling of a MEMS microsensor based on electrodynamic transduction, using both analytic and finite element analysis. Two coaxial planar inductors, external one and internal one, are used in the proposed microsensor design. Initially, we perform a modal analysis in order to determine the diaphragm resonant frequency and dynamic displacements for different diaphragm thicknesses. Next, the total sensitivity is deduced by coupling all domains involved in the microsensor operation, by using a lumped element model. We did build a model in order to study the microsensor sensitivity to predict the dynamic performance for different resonance frequency. The model shows that the best sensitivity is obtained around the resonant frequency, with a value estimated to some mV/Pa for the audible frequencies, and decreases to some μV/Pa for ultrasonic frequencies. This result proves that the undamped microsensor can reach a high sensitivity within a very narrow bandwidth.

Keywords: Micro-Electro-Mechanical Systems, Electrodynamic Transducer, Acoustic-Electrical-Mechanical Analogy, FEM simulation, Brownian noise, Diaphragm dynamic performance.

1 Introduction

Micro-Electro-Mechanical Systems, or MEMS, are generally considered as micro devices consisting of combination of micro mechanical sensors and/or actuators in addition to microelectronic circuits [1]. MEMS-based sensors have been under steady development for the last several decades, and a variety of devices have been developed aiming numerous applications like accelerometers [2], gyroscopes [3], Magnetometers [4] and acoustic applications [5–6]. Acoustic microsystem devices have been focused by many teams in order to achieve various kinds of specific results. The main objectives of acoustic MEMS research have been the miniaturization of existing acoustic devices to a microscopic scale, saving cost, and to obtain acoustic sensing devices with higher frequency range, which is difficult to achieve with the traditional ones [5]. Acoustic MEMS-based devices can be divided into two main domains: low frequency

M. Hadj Said, F. Tounsi, B. Mezghani and M. Masmoudi: University of Sfax, National Engineering School of Sfax, EMC research Group, Sfax, Tunisia, e-mails: mohamed.haj.said@gmail.com, fares.tounsi@isimsf.rnu.tn, brahim.mezghani@enis.tn, mohamed.masmoudi@enis.tn.
L. Rufer: University Grenoble Alpes, TIMA Laboratory, Grenoble, France, e-mails: Libor.Rufer@univ-grenoble-alpes.fr.

De Gruyter Oldenbourg, ASSD – Advances in Systems, Signals and Devices, Volume 6, 2018, pp. 35–52.
https://doi.org/10.1515/9783110448375-003

acoustic MEMS, such as microphones and microspeakers [7–8], and high frequency acoustic MEMS (like surface acoustic waves (SAW) devices and bulk acoustic waves (BAW) devices as oscillators and filtering components) [9–10]. Those devices use a variety of transduction or actuation techniques to transform the acoustic pressure wave into electrical signal or inversely. Main used methods are piezoelectricity, piezoresistivity, electrostatic and electromagnetic methods [8, 11–13]. In previous papers, a MEMS approach using the electrodynamic transduction was presented for detecting acoustic pressure variation [15]. The electrodynamic transduction was realized by two planar concentric inductors, which is different from the conventional method based on a permanent magnet to generate the B-field. The objectives of the current study are to determine the performance of the electrodynamic transduction for different acoustic applications.

The paper is organized as follows: after the operating principle description of the microsensor, we will present a mechanical modeling of the vibrating part using both analytical and FEM analysis of the micromachined structure. This will be performed for different diaphragm thicknesses. This section objective is to determine the mechanical properties such as the resonant frequency and the diaphragm displacement magnitude. The modeling will include the diaphragm dimensions optimization in order to achieve the targeted microsensor's dynamic performances in accordance with the available manufacturing technology. In the last section, the sensitivity will be evaluated by finding the coupling schemes between the involved transduction and the parameters of each transducer (acoustic -mechanical - electromagnetic).

2 Microsensor operating principle

The electrodynamic microsensor under study consists of two coaxial, concentric planar inductors (see Fig. 1); a fixed outer inductor B1 is placed on the substrate top surface and an inner inductor B2 is implemented on a square diaphragm suspended over a micromachined cavity. The diaphragm is composed of different materials stacks, which are fixed by the used technology. A DC bias current in B1, leads to the generation of a permanent magnetic field within B2. The vertical movement of the inner inductance B2 (same as of the diaphragm) within the created magnetic field will generate, at its ends, an induced output voltage proportional to the incident acoustic wave amplitude. In a previous work, both inductors sections and their mutual spacing were optimized to increase the magnetic field [14–15].

In the next section, the resonant frequency and displacement of the diaphragm will be deduced using the specific fabrication process [16].

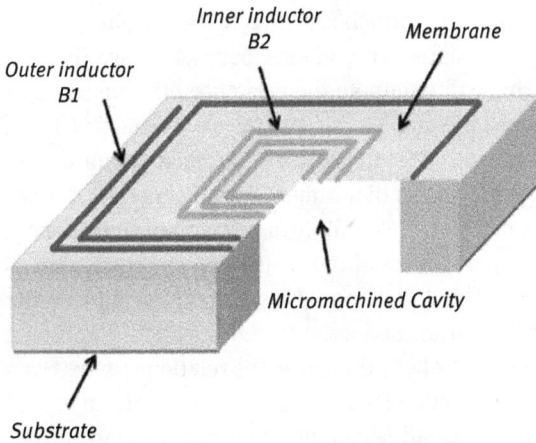

Fig. 1. 3D basic representation of the electrodynamic microphone structure.

3 Microsensor mechanical modeling

3.1 Resonance frequencies evaluation

Suspended structures are usually represented by a second order mechanical system represented by a mass with a damping spring. This linear model gives good results as the fluctuation has small amplitudes. The dynamic response of such a system is the solution of the following differential equation:

$$m \frac{d^2 x}{dt^2} + b \frac{dx}{dt} + kx = f_{ext} \tag{1}$$

where m is the effective mass of the suspended structure, k is the spring constant of the suspension, b is the damping coefficient (mechanical resistance), x is the mass displacement and f_{ext} is the sum of the external forces applied to the structure (corresponding to sound pressure in the microphone case). Using the Laplace transform, the frequency response is expressed by this equation:

$$X(j\omega) = \frac{1}{k} \frac{1}{1 - \left(\frac{\omega}{\omega_0}\right)^2 + j \frac{1}{Q} \frac{d\omega}{\omega_0}} \tag{2}$$

where: ω_0 is the resonant angular frequency and Q defines the quality factor of the vibrating structure. The term ζ is defined as the damping factor of the system. with:

$$\omega_0 = \sqrt{\frac{k}{m}}, \quad Q = \frac{k}{\omega_0 b} = \frac{1}{b}\sqrt{km}, \quad \zeta = \frac{b}{2\sqrt{km}} = \frac{1}{2Q}$$

The sensor diaphragm was designed to be attached through its periphery using a back-side bulk micromachining technique. This choice permits to avoid using openings around the diaphragm. This will eliminate the existence of acoustic short paths toward low frequencies, between the ambient air and the cavity underneath in order to widen the microsensor bandwidth [17–18]. So, the main microsensor design key parameter is the resonant frequency of the diaphragm, which generally defines the targeted frequency range, which can be either audible or ultrasonic. The resonant frequency is fixed by the geometrical parameters of the suspended structure such as size, thickness and attachments type. This latter should be well optimized in order to get an accurate behavior of the device in modal mode. An attached diaphragm structure has a vibrating natural frequency of ω_0, therefore, the relationship between the vibration amplitude and the system frequency has a peak at ω_0. This latter will have to be adjusted according to the targeted sensitivity dynamic response of our electrodynamic transducer depending on the application were it is intended to be used in. Inversely, for a squared diaphragm, attached at its periphery, the resonant frequency is defined by the following equation [19]:

$$f_1 = \frac{35.99}{2\pi L^2} \sqrt{\frac{D}{\rho t_h}}$$
(3)

where L is the diaphragm side length, ρ is the equivalent stacked material density, t_h is the diaphragm's elastic thickness and D is the flexural rigidity, given by:

$$D = \frac{E t_h^3}{12(1 - \upsilon^2)}$$
(4)

where E is the Young's modulus of the equivalent stacked materials, and υ its equivalent Poisson's ratio. Diaphragm thickness is a technological parameter, which is determined by the used fabrication process. Because different materials are deposited on the diaphragm top surface, some adjustment correction should be introduced to equation (3) and equation (4), especially its average mechanical proprieties such as Young's modulus and density, given by:

$$E_{moy} = \frac{\sum\limits_{i=1}^{n} E_i t_i}{\sum\limits_{i=1}^{n} t_i}, \quad \rho_{moy} = \frac{\sum\limits_{i=1}^{n} \rho_i t_i}{\sum\limits_{i=1}^{n} t_i}$$
(5)

where n is the number of diaphragm layers and the index i reflects the specific layer number forming the diaphragm. Next, the diaphragm resonance frequency as a function of its length, L, for different diaphragm thicknesses, is plotted in Fig. 2. To be able to clearly compare the results, we use three different diaphragm thicknesses, which produce three resonant frequencies: the first is inside the audible band (20 Hz to 20 kHz), the second is at the upper limit of the audible band and the third resonant frequency is at the lower limit of the ultrasound range. Using equation (3) with a diaphragm length of 1500 μm, resonant frequencies values were evaluated for different thicknesses. Results are summarized in Table 1.

Tab. 1. Resonant frequency values for increasing diaphragm thicknesses.

Diaphragm thickness [μm]	Resonance frequency [kHz]
0.3μm	1.619
3μm	17.48
10μm	46.28

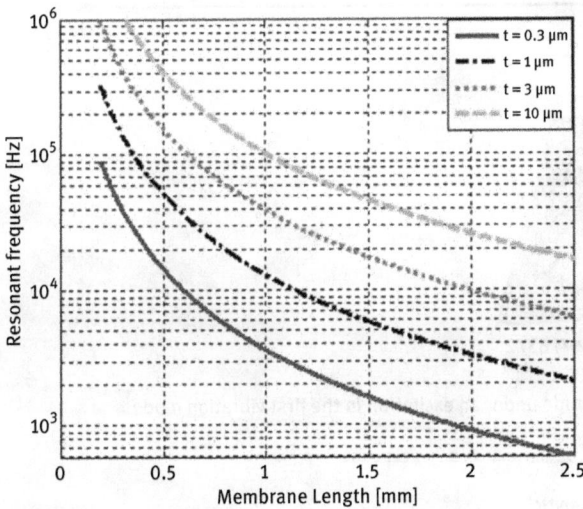

Fig. 2. Evaluation of the analytical diaphragm resonance frequency as a function of its side length L and thickness t.

On the other hand, Fig. 2 shows that the diaphragm length can be, as well, decreased in order to attain the same resonant frequency instead of increasing the thickness; however this solution will reduce the B-field produced by the outer inductance. For the adopted diaphragm, the mechanical equivalent effective mass and stiffness are, respectively, given by:

$$M_{dia}^{mec} = 0.607904 \rho \, t_h \, L^2 \tag{6}$$

$$K_{dia}^{mec} = 787.402 \frac{D}{L^2} \tag{7}$$

To validate the analytic resonant frequency, a modal analysis was performed using «Solid mechanics» module under Comsol©Multiphysics software. The simulation has been done on a diaphragm with a side length of 1500 μm and thickness of 0.3 μm (0.2 μm of silicon dioxide and 0.1 μm of silicon nitride). Both the simulation and the analytic values of the diaphragm properties are given in Table 2. We can notice that the differences between both results are slightly different due to meshing size. Figure 3 shows the diaphragm displacement magnitude under the first vibration mode.

Tab. 2. Evaluation of analytic and FEM mechanical properties of the square diaphragm with 1500 μm side length and 0.3 μm thikness.

Diaphragm properties	Analytic	FEM	Difference (%)
Resonance frequency (f_r) [kHz]	1.619	1.632	0.80
Effective mecanical mass (M_{dia}) [g]	1.025×10^{-9}	1.048×10^{-9}	2.19
Mecanical spring constant (K_{dia}) [N/m]	0.106	0.110	3.63

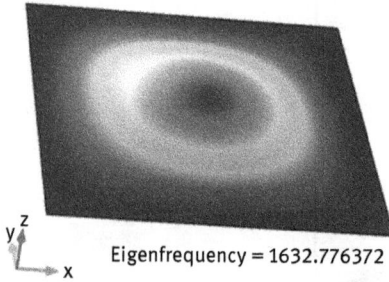

Eigenfrequency = 1632.776372

Fig. 3. Diaphragm deflection magnitude under an excitation in the first vibration mode.

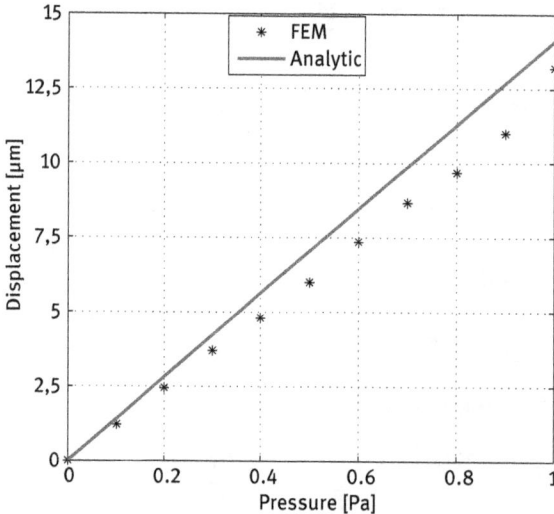

Fig. 4. Static diaphragm displacement as a function of the incident pressure.

3.2 Dynamic diaphragm displacement evaluation

The maximum displacement of a square diaphragm, under a static applied pressure Pin, is located in the center, and can be expressed by the following expression [1]:

$$\delta_{static} = P_{in} \frac{L^4}{1184 D} \tag{8}$$

In Fig. 4, we plot both FEM and analytic values of the diaphragm static displacement for different applied incident pressure magnitudes. Again, this was done using a diaphragm side of 1500 μm and thickness of 0.3 μm.

Since the diaphragm displacement is related to the frequency of the incident pressure wave, a harmonic simulation was performed using the «Shell» module under Comsol ® Multiphysics to obtain the dynamic diaphragm displacement. By applying a periodic pressure of 0.1 Pa and under a damping factor equal to 0.1, the frequency response of the plate was obtained using a frequency sweep in the resonance neighborhood of the obtained analytic values found in Table 1. In our study, we assume the simplest case in which the acoustic wave is purely sinusoidal with fixed amplitude. COMSOL simulation of the displacement in the diaphragm center as a function of the frequency is plotted in Fig. 5, and confirmed theoretically using equation (2). We can note that the displacement is optimum and found around the already set resonant frequencies previously evaluated (which are 1.6 kHz, 17.5 kHz and 46.3 kHz).

4 Microsensor sensitivity evaluation

4.1 Electro-acoustic modeling theory and basis

To be able to estimate the microsensor sensitivity, we need to extract an equivalent lumped element model imitating the physical device behavior. The lumped model should gather mechanical, acoustic and electromagnetic phenomena excitation. To this end, an analogy between different involved physical domains should be used. This will ease the solution of mechanical and/or acoustic vibrating systems through converting these coupled problems to the corresponding electrical domain and then solving the resulting electrical circuit using classical theory. The acoustic elements encountered in the model involve resistance, mass or compliance according to the edges of the modeled tube and the way in which it behaves in response to an acoustic flow (see Table 3). In such a model, the acoustic pressure corresponds to a voltage; an acoustic flow corresponds to a current and volume displacement to an electrical charge.

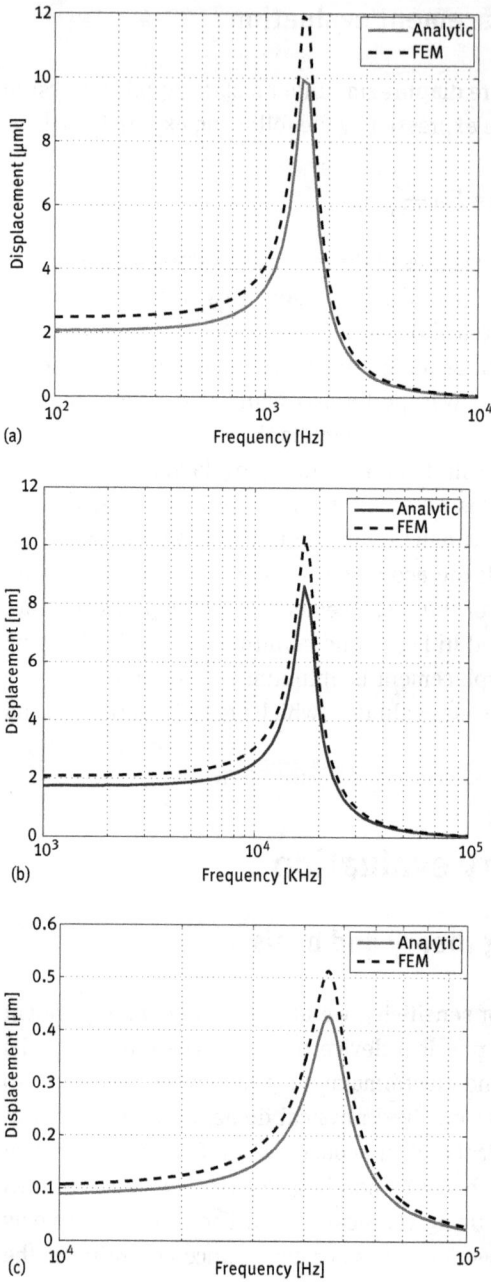

Fig. 5. Diaphragm spectrum displacement as a function of the frequency, for a diaphragm length of 1500 μm and thickness equal to (a) t_h =0.3 μm, (b) t_h =3 μm, (c) t_h =10 μm.

Tab. 3. Conversion table between Acoustic/Electrical/Mechanical domains.

Acoustic Parameter	Unit	Electrical equivalent parameter	Unit	Mechanical equivalent Parameter	Unit
Pressure, p	$Pa = N/m^2$	Voltage, U	V	Force, F	N
Flow, w	m^3/s	Current, I	A	Velocity, v	m/s
Displacement, z	m^3	Charge, Q	C [As]	Displacement, x	m
Compliance, C_a	$m^3/Pa = m^5/N$	Capacitance, C_e	F	Compliance, C_m	m/N
Stiffness K_a	$Pa/m^3 = N/m^5$	Capacitance^{-1}	F^{-1}	Stiffness	N/m
Acoustic mass, M_a	$kg/m^4 = Ns^2/m^5$	Inductance, L_e	H	Mass, M_m	kg
Acoustic resistance, R_a	$Pa\ s/m^3 = Ns/m^5$	Resistance, R_e	Ω	Resistance, R_m	Ns/m

When applying analogy between different energy fields, a lumped element model of the microsensors can be proposed. This analogy consists in connecting in series all elements crossed by the same acoustic flow and in parallel elements corresponding to a flow addition. From the lumped element modeling, we can determine the sensitivity, the frequency response (in particular amplitude and phase), the complex acoustic impedance of the diaphragm, the electrical impedance and the intrinsic electrical noise of the microsensors. In next section, different domains in the microsensors will be analyzed and discussed to derive the final equivalent lumped model.

4.2 Microsensors equivalent electro-acoustic circuit

The sensitivity depends on the domains involved in the microsensor operation principle (acoustic-mechanical-magnetic). Initially, in the acoustic domain when the pressure hits the diaphragm surface, an acoustic wave will radiate outward and can be modeled with radiation impedance by an acoustic resistance and masses given by [17]:

$$\underline{Z_{rad}^{ac}} = \frac{1}{8}\frac{\rho_{air}}{C_{air}}\omega^2 - j\frac{4}{3\pi}\frac{\rho_{air}}{L}\omega = R_{rad} + jM_{rad}\omega \tag{9}$$

where ρ_{air} is the air density and C_{air} is the sound velocity in the air. On the other hand, when the incident pressure Pin physically hits the diaphragm top surface, this latter fluctuates. Therefore, its behavior can be modeled by a mechanical stiffness and mass, when neglecting the axial stress occurring in the fabrication process, which is given by:

$$\underline{Z_{mem}^{me}} = j\omega M_{dia}^{mec} + \frac{K_{dia}^{mec}}{j\omega} \tag{10}$$

The diaphragm fluctuation due to the acoustic pressure will apply a pressure in the gap underneath the diaphragm, P_{cav}. Since the air, considered as a perfect gas, in the micromachined cavity volume, closed at its end, is compressible the acoustic impedance of the cavity can be assimilated by:

$$\underline{Z^{ac}_{back}} = \frac{\rho_{air}C^2_{air}}{j\omega V} = \frac{K^{ac}_{cav}}{j\omega} \tag{11}$$

where V is the volume of the back cavity which equivalent weight is neglected in our model. This assumption is valid if the cavity side length is bounded between $\frac{1}{20}\sqrt{\frac{\pi}{f}}$ and $\frac{10\sqrt{\pi}}{f}$. In the microsensor support, we need to add a vent to equalize the back cavity pressure to the atmospheric pressure. This slit is equivalent to a rectangular hole, which is equivalent to an acoustic resistance, with b as a length; w_h is the width and l_h the thickness. The acoustic resistance and mass for rectangular capillary slots are given by [17]:

$$\underline{Z^{ac}_h} = \frac{12\eta_{air}l_h}{b^3 w_h} + j\omega \frac{12\rho_{air}l_h}{b w_h} = R_{hole}ac + jM^{ac}_{hole}\omega \tag{12}$$

Finally, the diaphragm displacement will generate an induced voltage at the ends of the inner inductor, so a last electro-magnetic phenomenon will be considered and included, which is the magnetic induction link, causing the induced voltage.

The different parameters introduced above can be gathered in a lumped element model representing all the previously explained effects (see Fig. 6). The developed circuit relates the different domains through the transformers and the gyrator with appropriate coupling coefficients. The coupling coefficient between mechanical and acoustic domains is S, which represents the diaphragm surface. In fact, this coefficient will relate also the flow rate w and the velocity of the diaphragm v as shown:

$$F = S P_{in}, \quad w = S v \tag{13}$$

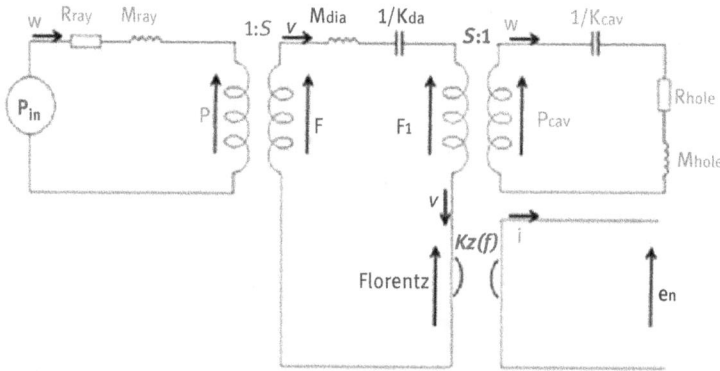

Fig. 6. Lumped element model showing different coupling domains of the microsensor.

On the other hand, the coefficient K is defined as the coupling gain between the magnetic and the mechanic domain and is given by [20]:

$$K_z(f) = n_1 n_2 A_2 I_1 \xi(f) \tag{14}$$

where n_1 and n_2 are the number of turns of the inner and outer inductors, $\xi(f)$ is the out-of-plane diaphragm displacement as a function of the frequency given in equation (2), I_1 is the current flowing in the outer inductor and A_2 is a geometric constant depending on the average distance between the two inductors ε_a and the coil side a is given by:

$$A_2 = \frac{2\mu_0}{\pi\varepsilon_a^2} \left(\sqrt{\varepsilon_a^2 + (a - \varepsilon_a^2)} - \sqrt{2}\varepsilon_a \right) \tag{15}$$

This coefficient relates the force $F_{Lorentz}$ produced to the induced current i_2, and relates the induced voltage with the mechanical velocity of the diaphragm as given:

$$F_{Lorentz} = K_z(f) i_2 , \quad e = K_z(f) v \tag{16}$$

To simplify the microsensor model, we have used the relations between the different parameters on each side of the transformers and the gyrator. Generally, the mechanical part is represented by a force F and an impedance Z^m, while the acoustic part, by an acoustic pressure p and an impedance Z^a, as shown in Fig. 7.a. In Fig. 7.b, the acoustic equivalent circuit without transformer is represented, where the primary is brought to the secondary. It contains the acoustic equivalent impedances and source given by $Z^{am} = Z^m/S^2$ and $p_F = F/S$. The contrary will be done for the mechanical equivalent circuit where the acoustic part will be brought to the mechanical part (see Fig. 7.c).

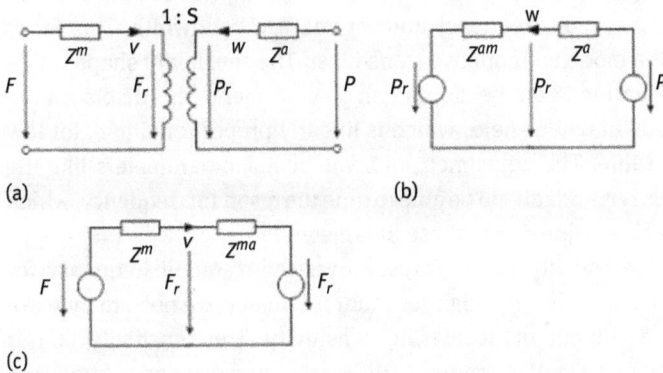

(a)

(b)

(c)

Fig. 7. Representation of the mechanical-acoustic coupling in the form of (a) coupling scheme (b) equivalent acoustic scheme (c) equivalent mechanical scheme.

4.3 Frequency response and sensitivity

After simplifying the circuits by bringing all the components in the acoustic part, and using the Kirchhoff's laws, the acoustic flow can be given by:

$$w = \frac{S^2 P_{in} P}{P^2 \left[S^2 (M_{ray}^{ac} + M_{hole}^{ac}) + M_{dia}^{mec} \right] + PS^2 \left[R_{ray}^{ac} + R_{hole}^{ac} \right] + \left[K_{dia}^{mec} + S^2 K_{cav}^{ac} \right]} \tag{17}$$

In our model, we assume that the force F due to the incident pressure is higher compared to the electromagnetic force. Finally, when using equation (13) and equation (16), the relation between the induced voltages and the applied pressure of the sensor can be given by:

$$f(p = j\omega) = \frac{e_{in}}{P_{in}}$$

$$= \frac{SK_z(f)P}{P^2 \left[S^2 (M_{ray}^{ac} + M_{hole}^{ac}) + M_{dia}^{mec} \right] + PS^2 \left[R_{ray}^{ac} + R_{hole}^{ac} \right] + \left[K_{dia}^{mec} + S^2 K_{cav}^{ac} \right]} \tag{18}$$

From equation (18), we can notice that the transfer function between the induced voltage and the applied pressure increases by increasing the coefficient $K_z(f)$, especially the current, and the diaphragm displacement. The overall sensitivity of the microsensor can be deduced:

$$S_{en} = \frac{\Delta e_{in}}{\Delta P_{in}} \tag{19}$$

The sensitivity was drawn in Fig. 8 as a function of frequency for different diaphragm thicknesses, while keeping the same diaphragm length 1500 μm.

We can observe that for the three structures, the sensitivity has a narrow bandwidth with a maximum at the diaphragm resonant frequency. So, we can deduce that our microsensor is almost a mono-frequency detector. It can be used to produce high sensitivity around a specific frequency and very narrow bandwidth. This result is very interesting for the modeled inductive transducer. The sensitivity shape is due to the dependence of equation 18 on the diaphragm displacement. This displacement originates from the radial magnetic field, which is linearly proportional to ζ, for low amplitude fluctuation value. The adjustment of some acoustic parameters like the cavity stiffness and mass is mandatory to be able to tune the resonant frequency, which can shift from its original position due to these parameters.

Figure 9 shows the sensitivity variation peak over the resonant frequency for different diaphragm thickness. To set a high resonant frequency, we need to increase the thickness on the detriment of decreasing sensitivity. The sensitivity of our microsensor can drop from 2.23 mV to around 6 μV/Pa when the resonance frequency is set around 47 kHz. The obtained sensitivity performances make the design more useful in applications like acoustic emission sensors and ultrasonic testing sensors, which require a high sensitivity within a narrow bandwidth (resonance model) [21–22].

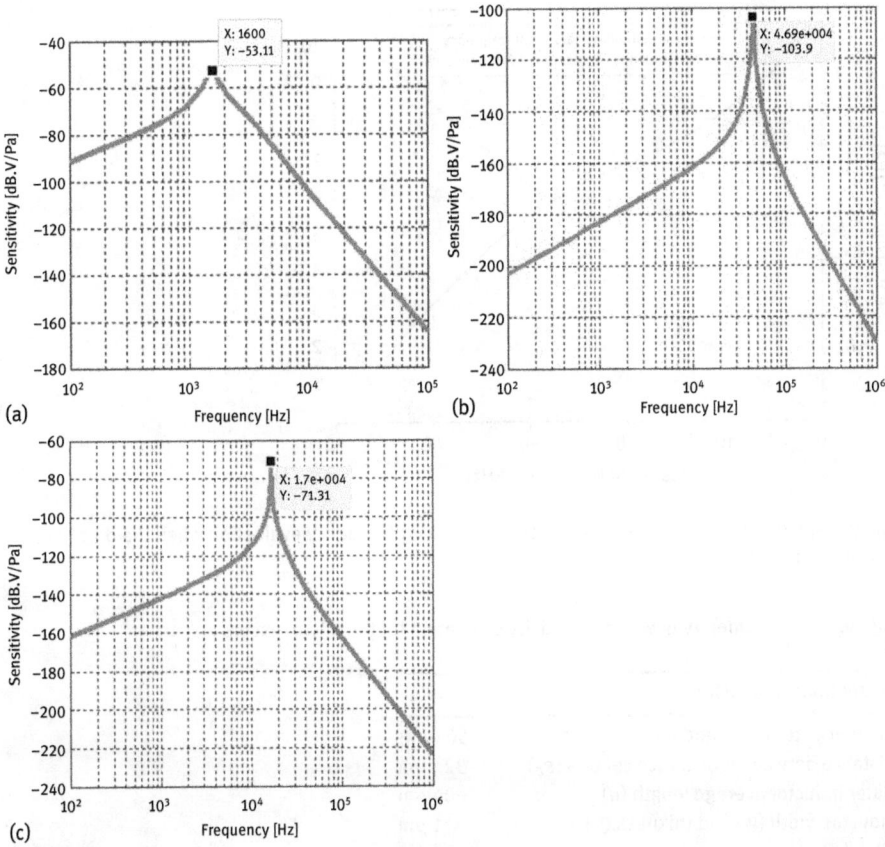

Fig. 8. Evaluated electrodynamic microsensor sensitivity as a function of frequency for a diaphragm length of 1500 μm and thicknesses equal to (a) t_h =0.3 μm, (b) t_h =3 μm, (c) t_h =10 μm.

In absence of damping, the electrodynamic microsensor is able to provide a very high sensitivity around the mechanical resonance frequency. The sensitivity is three orders of magnitude higher than in case of damped case of the electrodynamic microphone [23]. So, we can confirm that enlarging the microsensor bandwidth will reduce significantly the overall sensitivity and vice versa. This result makes our transducer competitive with the traditional ultrasonic counterpart found in literature in term of sensitivity [13, 17]. The final dimension and the different parameters used for our model are shown in Tab. 4.

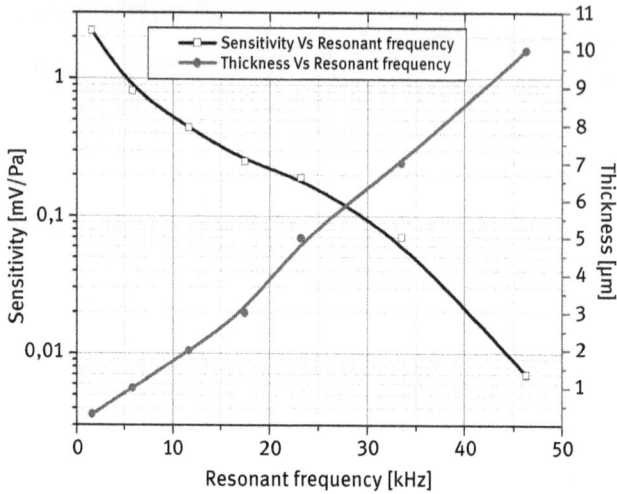

Fig. 9. Sensitivity and thicknesses for different resonant frequency evaluation when the diaphragm side equal to 1500 μm.

Tab. 4. Main parameters used for sensitivity evaluation for the microsensors.

Microphone dimension	Value
Number of turns, n_1 and n_2	50 turns
Distance between inductance centers (ε_a)	324 μm
Outer inductor average length (a)	1604 μm
Inductor width (w) and thikness (t_i)	1 μm
Pitch between spirals (s)	1 μm
Diaphragm side (L)	1500 μm
Gap thickness (substrate thickness) (h)	675 μm
Hole length (l_h)	200 μm
Hole width (w_h) and side (b)	50 μm

5 Conclusion

In this paper, we performed both modal and harmonic modeling for an electrodynamic MEMS microsensor over a large bandwidth. Firstly, the resonant frequency was deduced for different possible diaphragm thicknesses, using both analytic and FEM simulation. Thereafter, the static and harmonic displacements were deduced showing an optimum around the chosen resonant frequency of the diaphragm. Finally, the dynamic sensitivity of the microsensor was found by gathering all parameters from involved domains. We can conclude that the adopted electrodynamic transducer design gives higher sensitivity around the resonant frequency but within a narrow bandwidth. This result is due to the dependence of the sensitivity to the diaphragm displacement originating from electromagnetic coupling coefficient that is linearly proportional to the diaphragm displacement. The microsensor sensitivity is around few mV/Pa for small resonant frequencies and drops down to μV/Pa for ultrasonic frequencies.

Bibliography

[1] B. Minhang. *Analysis and Design Principles of MEMS Devices*. Elsevier B.V, First Edition 2005, :104–105, 2005.

[2] T.A. Roessig, R.T. Howe, A.P. Pisano and J.H. Smith. Surface micromachining resonant accelerometer. *Int. Conf. on solid state Sensors and Actuators*, Chicago, 2:859–862, 1997.

[3] Y. Oh, B. Lee, S. Baek, H. Kim, J. Kim, S. Kang and C. Song. Surface- micromachined tunable vibratory gyroscope. in: Proceeding of the IEEE International Microelectromechanical Systems Workshop, Nagoya, Japan :272–277, 1997.

[4] M.H. Said, F. Tounsi, M. Masmoudi, P. Gkotsis and L. Francis. A MEMS resonant magnetometer based on capacitive detection. *Int. Multi-Conf. on Systems, Signals and Devices*, Hammamet, Tunisia, 2013.

[5] V. Magori and H. Walker. Ultrasonic Presence Sensors with Wide Range and High Local Resolution. Ultrasonics, Ferroelectrics and Frequency Control. *IEEE Trans. on Ultrasonics Ferroelectrics and Frequency Control*, 34:202–211, 1987.

[6] J.J. Neumann Jr, K.J. Gabriel. CMOS-MEMS membrane for audio-frequency acoustic actuation. *Sensors and Actuators A*, 95:175–182, 2002.

[7] B.A. Ganji and B.Y. Majlis. Design and fabrication of a new MEMS capacitive microphone using a perforated aluminum diaphragm. *Journal of Sensors and Actuators A*, 149:29–37, 2009.

[8] G. Sugandi and B.Y. Majlis. Fabrication of MEMS Based Microspeaker Using Bulk Micromachining Technique. *Advanced Materials Research*, 254:171–174, 2011.

[9] T. Mattila, J. Kiihamäki, T. Lamminmäki, O. Jaakkola, P. Rantakari, A. Oja, H. Seppä, H. Kattelus and I. Tittonen. A 12 MHz micromechanical bulk acoustic mode oscillator. *Sensors and Actuators A*, 101:1–9, 2002.

[10] A. Zaki, H. Elsimary and M. Zaghloul. Miniature SAW device using MEMS technology. *Microelectronics Journal*, 38:426–429, 2007.

[11] S. Horowitz, T. Nishida and L. Cattafesta. Development of a micromachined piezoelectric microphone for aeroacoustics applications. *Journal of Acoustical Society of America*, 122:3428–3436, 2007.

[12] D. P. Arnold, S. Gururaj, S. Bhardwaj, T. Nishida and M. Sheplak. A piezoresistive microphone for aeroacoustic measurements. *Int. Mechanical Engineering Congress and Exposition*, :281–288, 2001.

[13] S. T. Hansen, A. S. Ergun, W. Liou, B. A. Auld and B. T. Khuri-Yakub. Wideband micromachined capacitive microphones with radio frequency detection. *Journal of the Acoustical Society of America*, 116:828–842, August 2004.

[14] M. H. Said, G.S. Sandeep, F. Tounsi, B. Mezghani, M. Masmoudi and V.R. Rao. Numerical Magnetic Analysis for a Monolithic Micromachined Electrodynamic Microphone. *ICMEMS Conference*, IIT MADRAS-India, 2014.

[15] F. Tounsi, L. Rufer, B. Mezghani, M. Masmoudi and S. Mir. Highly Flexible Membrane Systems for Micromachined Microphones-Modeling and Simulation. 3rd *IEEE SCS'*09, Djerba, Tunisia, 2009.

[16] http://www.cen.iitb.ac.in/cen/index.php

[17] J. Esteves, L. Rufer, S. Basrour and D. Ekeom. CMOS-MEMS technology with front-end surface etching of sacrificial SiO2 dedicated for acoustic devices. 5th *IEEE Int. Workshop on Advances in Sensors and Interfaces* (IWASI'13), Bari, Italy, :154–159, June 2013.

[18] Z. Zhou, L. Rufer and M. Wong. Damped Aero-Acoustic Microphone with Improved High-Frequency Characteristics. *IEEE Microelectromechanical Systems*, 23(5):1094–1100, 2014.

[19] L. Rufer, C. Domingues, S. Mir, V. Petrini, J.-C. Jeannot and P. Delobelle. A CMOS compatible ultrasonic transducer fabricated with deep reactive ion etching. *IEEE Journal of Microelectromechanical Systems*, 15(6):1766–1776, 2006.

[20] F. Tounsi. Theoretical Electromagnetic survey: Application to a planar CMOS-MEMS electrodynamics microphone. *Novel Advances in Microsystems Technologies and Their Applications*. :205–246, CRC Press (Taylor & Francis), 2013.

[21] S. Huang, M. Li, Y. Xu, D. Xu, X. Xie and X. Cheng. Research on Embedded Sensors for Concrete Health Monitoring Based on Ultrasonic Testing. 4th *Int. Conf. on the Durability of Concrete Structures*, Purdue University, West Lafayette, IN, USA, :24–26 July 2014.

[22] J. Posada-Roman, J.A. Garcia-Souto and J. Rubio-Serrano. Fiber Optic Sensor for Acoustic Detection of Partial Discharges in Oil-Paper Insulated Electrical Systems. *Sensors*, 12:4793–4802, 2012.

[23] M. Hadj Said, F. Tounsi, S.G. Surya, B. Mezghani, M. Masmoudi and V.R. Rao. A MEMS-based Shifted Membrane Electrodynamic Microsensor for Microphone Applications. *Journal of Vibration and Control*, February 2016.

Biographies

Mohamed Hadj Said was born in Monastir, Tunisia in 1987. He received the B.Sc.'10 and M.Sc'12 degrees both from the Institut Supérieur d'Informatique et de Mathématiques de Monastir (ISIMM, Tunisia). His Master is done in collaboration with the laboratory of Sensors, Microsystems and Actuators of Louvain-la-Neuve (SMALL), Belgium. He is currently a PhD student, in the National Engineering School of Sfax (ENIS), working on the design, modeling and simulation of a novel MEMS microphone based in magnetic detection in CMOS Technology, in collaboration with TIMA Laboratory France and IIT Bombay India. Furthermore, he interests in magnetic sensors, surface acoustic wave modeling sensors and acoustics sensors.

Fares Tounsi received the B.Sc.'01 and M.Sc.'03 degrees from National Engineering School of Sfax (ENIS) in Tunisia, and the Ph.D.'10 in Micro and Nano-electronics from Grenoble Institute of Technology (INPG), France. He was a visiting Professor during 2013 in the Berkeley Sensor & Actuator Center (UC Berkeley, USA). He is actually an assistant Professor in the Institut Supérieur d'Informatique et de Mathématiques de Monastir (ISIMM), Tunisia. Currently, he is working on the design, modeling and characterization of new CMOS-compatible micromachined sensors and actuators. Specifically, he is interested in novel designs of microphones, inertial sensors and RF switches. In addition, he is now focusing on the field of nanotransducers, evaluation of new materials/structures for MEMS and advanced microsystems.

Libor Rufer received the Engineer and Ph.D. degrees from Czech Technical University, Prague, Czech Republic. Until 1993, he was with the Faculty of Electrical Engineering, Czech Technical University, Prague. Since 1994, he has been an Associate Professor, and later a Senior Scientist with the University of Grenoble, France. In 1998, he joined the Microsystems Group of TIMA Laboratory. His expertise is mainly in MEMS-based sensors and actuators, electroacoustic and electromechanical transducers and their applications in acoustics, ultrasonics, and energy harvesting.

Brahim Mezghani received the Bachelor and Master of Science degrees from the University of Minnesota, USA. This was on low temperature noise characterization of HIGFETs with Honeywell's sensors division in Minneapolis, Minnesota. He received the Ph.D. and HDR degrees from National Engineering School of Sfax (ENIS) in Tunisia, where he is currently holding an associate professor position at the department of Electrical Engineering. His PhD research activities were concentrated on the development of a new CMOS MEMS acoustic sensor structure. His HDR research work has been mainly conducted on new 3D analytic and numerical analysis of CMOS MEMS convective accelerometers. Dr. Mezghani has currently several ongoing joint research activites in collaboration with research groups from the LIRMM, GeePS and TIMA Labs in France, SMALL from the UCL in Belgium and IIT-B in India. Dr. Mezghani is currently with the METS research unit at the ENIS where he is supervising the research group Micro and Nano-Systems design since 2008. The focus of his research interests is on design, modeling and simulation of micromachined sensor structures. This includes both the mechanical micromachined part and its processing low noise microelectronics. His interests include also the development of nanomaterials for sensor applications and for performance enhancement of MEMS sensors.

Mohamed Masmoudi was born in Sfax, Tunisia, in 1961. He received the Engineer in electrical Engineering degree from the National Engineers School of Sfax, Sfax, Tunisia in 1985 and the PhD degree in Microelectronics from the Laboratory of Computer Sciences, Robotics and Microelectronics of Montpellier, Montpellier, France in 1989. From 1989 to 1994, he was an Associate Professor with the National Engineers School of Monastir, Monastir, Tunisia. Since 1995, he has been with the National Engineers School of Sfax, Tunisia, where, since 1999, he has been a Professor engaged in developing Microelectronics in the engineering program of the university, and where he is also the Head of the METS research unit. He is the author and co-author of several papers in the Microelectronic field. He has been a reviewer for several journals. He is the founder and Chair of IEEE Solid State Circuits Tunisia Chapter. Dr. Masmoudi organised several international Conferences and has served on several technical program committees.

T. Ettaghzouti, N. Hassen and K. Besbes

A Novel Multi-Input Single-Output Mixed-Mode Universal Filter Employing Second Generation Current Conveyor Circuit

Abstract: In this paper, a new multi-input single-output (MISO) mixed mode universal filter is proposed. This circuit is composed by four second generation current conveyor circuits (CCII), two grounded capacitors and four resistances. It can provide all the five standard filter functions, namely low-pass, band-pass, high-pass, band-stop and all-pass by appropriately connecting the input terminals. The proposed filter maintains the following advantages: (i) three input terminals and one output terminal for each mode, (ii) the natural angular frequency (ω_0) and the quality factor (Q) can also be controlled independently (iii) low active and passive sensitivity performances. TSPICE simulation results, using $0.18\,\mu m$ TSMC CMOS technology with a supply voltage of $\pm 0.75\,V$, are included to verify the workability of the proposed filter. The given results are agreed well with the theoretical.

Keywords: Second generation current conveyor CCII, mixed mode, universal filter, MISO

1 Introduction

Analogue filters are considered as basic blocks of many electronic systems and applications as telecommunication systems, radio frequency applications, data conversion systems, instrumentation and control systems. This kind of circuit can be used to remove the superimposed noise on the analogue signal before it reaches the analogue to digital converter or to pass one or more desirable bands of frequencies and simultaneously reject one or more undesirable bands frequencies.

In the literature, many configurations of active universal filters using a current conveyor circuits such as second generation current conveyor (CCII), current controlled conveyor (CCCII), inverting second generation current conveyor (ICCII) and differential voltage current conveyor (DVCC) have received a considerable attention within reason of its good linearity, simple circuit, low power consumption and wide bandwidth. Taking into account the number of input and output ports, these filters can be classified into three groups. There are those who have multiple inputs and

T. Ettaghzouti, N. Hassen and K. Besbes: Micro-electronics and instrumentation laboratory University of Monastir Monastir, Tunisia, e-mails: thourayataghzouti@yahoo.fr, nejib.hassen@fsm.rnu.tn, kamel.besbes@fsm.rnu.tn.

K. Besbes: Centre for Research on Microelectronics and Nanotechnology of Sousse, Technopole of Sousse, Tunisia, e-mails: kamel.besbes@fsm.rnu.tn.

De Gruyter Oldenbourg, ASSD – Advances in Systems, Signals and Devices, Volume 6, 2018, pp. 53–64.
https://doi.org/10.1515/9783110448375-004

a single output terminal (MISO) [1–5], others who have single input and multiple output terminals (SIMO) [6–10] and others configurations who have multiple-input, multiple-output terminals (MIMO) [11–13] operating in current, voltage and mixed mode.

The use of several modules in the same system has forced the designers to utilize current and voltage mode filters in the same time. For that, several research works have been created in order to concatenate the both operating mode at the same circuit.

For example, a new mixed mode universal filter composed by five single output second generation current conveyor circuits, seven resistances and two grounded capacitors has been proposed in 2004 [14]. This circuit can realize all the five standard filter functions. It has three input terminals and one output terminal for each mode: It can be driven by voltage or current and its output can be voltage or current. In 2006, Neeta Pandey and al. have proposed a generalized mixed mode universal filter configuration that may be used in all possible mode, voltage mode, current mode, trans-impedance mode and trans-admittance mode by the control of two switches [15]. This novel architecture uses three single outputs CCII, three capacitors and four resistances: The factor quality and the natural angular frequency of this filter cannot be controlled independent.

In this work, a new mixed mode universal filter composed by four second generation current conveyor circuits, two grounded capacitors and four resistances is presented. This filter is characterized by three input and one output terminals for each mode. It can realize all the standard filter functions which are high-pass, band-pass, low-pass, notch and all-pass filters, by appropriately connecting the input terminals and without changing the circuit topology. All the simulation results are obtained by using TSPICE with TSMC 0.18 μ m CMOS process parameters.

2 Proposed circuit

The block diagram of second generation current conveyor circuit (CCII) is shown in Fig. 1. It has three terminal devices X, Y and Z. The input voltage applied to Y terminal is perfectly conveyed to X terminal and the input current applied to X terminal is conveyed to the Z terminal with the same direction if CCII is positive (CCII+) or reverse direction, if CCII is negative (CCII–).

In ideal case, the input- output current-voltage relationships between different terminals of CCII can be described by the following equations:

$$I_Y = 0 , \quad V_Y = V_X , \quad I_X = I_Z^+ = -I_Z^-$$

But in real circumstances, the CCII has parasitic elements as a parallel resistor with a capacitor on terminals Y and Z and a parasitic resistor RX on terminal X.

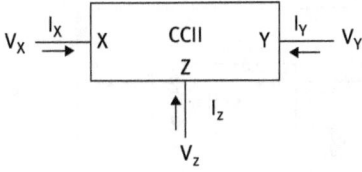

Fig. 1. Symbolic scheme of second generation current conveyor.

The characteristics equations are become as follows:

$$I_Y = \frac{V_Y}{R_Y \,//\, C_Y} \,, \quad V_Y = \beta V_X + R_X I_X \,,$$

$$I_X = \alpha I_Z^+ + \frac{V_Z^+}{R_Z \,//\, C_Z} = -\alpha I_Z^- + \frac{V_Z^-}{R_Z \,//\, C_Z}$$

where α and β denote respectively the current gain and the voltage gain of CCII.

The MOS implementation of second generation current conveyor circuit CCII [16] is presented in Fig. 2. It is composed by differential pair transistors (M1, M2) biased by a current source transistor M15, three current mirrors formed by three matched transistors (M3, M4, M5), (M6, M7, M8) as well as (M11, M12, M13) and voltage source follower composed by (M9, M10, M14).

A universal active filter has been implemented by connecting four prototypes of CCII presented in Fig. 2 two grounded capacitors and four resistances as shown in Fig 3. This filter can be operated either in voltage mode or in current mode by making appropriately the connection the input terminals.

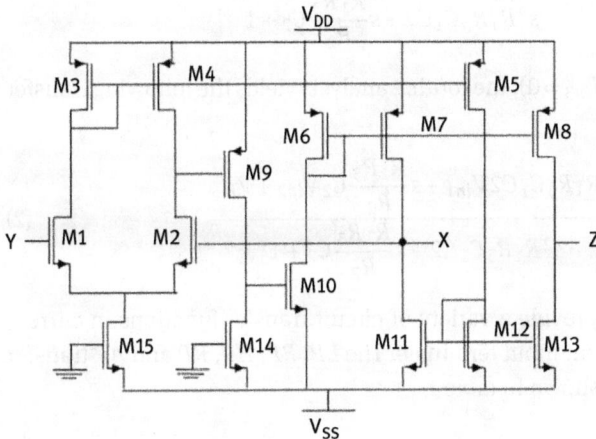

Fig. 2. Second generation current conveyor circuit CCII.

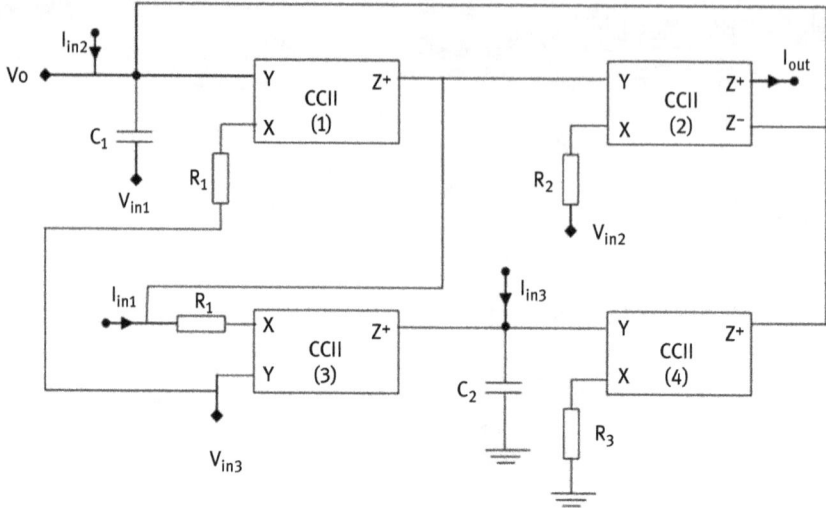

Fig. 3. Proposed multi-input signal-output mixed mode universal filter.

2.1 Analysis of ideal case

Considering the ideal case of second generation current conveyor circuit and when the input voltage terminals V_{in1}, V_{in2} and V_{in3} are connected to grounded ($V_{in1} = V_{in2} = V_{in3} = 0$), the routine analysis of proposed filter circuit yields the following transfer function:

$$I_{out} = \frac{R_1}{R_2} \times \frac{s^2 R_1 R_3 C_1 C2 I_{in1} + s R_3 C_2 I_{in2} + I_{in3}}{s^2 R_1 R_3 C_1 C2 + s \dfrac{R_1 R_3}{R_2} C_2 + 1} \tag{1}$$

In voltage mode ($I_{in1} = I_{in2} = I_{in3} = 0$), the routine analysis yields the following transfer function:

$$V_{out} = \frac{s^2 R_1 R_3 C_1 C2 V_{in1} + s \dfrac{R_1 R_3}{R_2} C_2 V_{in2} + V_{in3}}{s^2 R_1 R_3 C_1 C2 + s \dfrac{R_1 R_3}{R_2} C_2 + 1} \tag{2}$$

Equations (1) and (2) above provide a variety of circuit transfer functions in current and voltage mode with different input terminals. The LP, BP, HP, RP and AP transfer functions can be realized as shown in table 1.

Tab. 1. Input terminals values selection for each filter function responses.

	Voltage mode filter ($I_{in1} = I_{in2} = I_{in3}$)			Current mode filter ($V_{in1} = V_{in2} = V_{in3}$)		
	V_{in1}	V_{in2}	V_{in3}	I_{in1}	I_{in2}	I_{in3}
Low pass filter (LP)	0	0	1	0	0	1
Band pass filter (BP)	0	1	0	0	1	0
High pass (HP)	1	0	0	1	0	0
Reject pass (RP)	1	0	1	1	0	1
All pass (AP)	1	1	1	1	1	1

The pole frequency (ω_0), the quality factor (Q) and bandwidth (BW) $\dfrac{\omega_0}{Q}$ of each filter responses can be calculated as:

$$\omega_0 = \frac{1}{\sqrt{R_1 R_3 C_1 C_2}}$$

$$Q = R_2 \sqrt{\frac{C_1}{R_1 R_3 C_2}}$$

$$BW = \frac{1}{R_2 C_1}$$

It can be noted from (4) and (5) that the factor quality and bandwidth of proposed filter can be controlled independent of ω_0 by varied R_2.

2.2 Analysis of non-ideal case

The effects of non idealities of CCII can affect on the property of proposed filter (expressions of pole frequency, quality factor and bandwidth). The characteristic polynomial of current and voltage transfer functions have becomes:

$$I_{out} = \frac{R_1}{R_2} \times \frac{s^2 \alpha_1 \alpha_3 \alpha_4 \beta_1 \beta_4 R_1 R_3 C_1 C2 I_{in1} + s\alpha_3 \alpha_4 \beta_4 R_3 C_2 I_{in2} + \alpha_2 I_{in3}}{s^2 \alpha_1 \alpha_2 \alpha_3 \alpha_4 \beta_1 \beta_2 \beta_4 R_1 R_3 C_1 C2 + s\alpha_3 \alpha_4 \beta_4 \dfrac{R_1 R_3}{R_2} C_2 + \alpha_2 \beta_2} \tag{3}$$

$$V_{out} = \frac{\gamma\left(s^2 \alpha_2 C_1 V_{in1} + s\dfrac{1}{R_2} V_{in2}\right) + \left(\alpha_2 \beta_1 \beta_2 + s\alpha_3 \alpha_4 \beta_4 \dfrac{R_1 R_3}{R_2}(\beta_1 - \alpha_1) C_2\right) V_{in3}}{s^2 \alpha_1 \alpha_2 \alpha_3 \alpha_4 \beta_1 \beta_2 \beta_4 R_1 R_3 C_1 C2 + s\alpha_3 \alpha_4 \beta_4 \dfrac{R_1 R_3}{R_2} C_2 + \alpha_2 \beta_2} \tag{4}$$

with:

$$\gamma = \alpha_3 \alpha_4 \beta_2 \beta_3 \beta_4 R_1 R_3 C_1$$

The expressions of natural angular frequency, the quality factor and bandwidth of both modes have become as:

$$\omega_0 = \frac{1}{\sqrt{\alpha_1 \alpha_3 \alpha_4 \beta_1 \beta_4 R_1 R_3 C_1 C_2}} \tag{5}$$

$$Q = R_2 \sqrt{\frac{\alpha_1 \beta_1 C_1}{\alpha_3 \alpha_4 \beta_4 R_1 R_3 C_2}} \tag{6}$$

$$BW = \frac{1}{\alpha_1 \beta_1 R_2 C_1} \tag{7}$$

2.3 Active and passive sensitivities

Because of the tolerances in component values and the non-idealities of second generation conveyors CCIIs, the response of the actual assembled filter will deviate from the ideal response. As a mean for predicting such deviations, the filter designer employs the concept of sensitivity. The sensitivity function is defined as:

$$S_X^Y = \frac{X}{Y} \times \frac{\partial Y}{\partial X}$$

The active and passive sensitivities of proposed circuit can be found as

$$S_{\alpha_1 \alpha_3 \alpha_4 \beta_1 \beta_4}^{\omega_0} = -\frac{1}{2} , \ S_{R_1 R_3 C_1 C_2}^{\omega_0} = -\frac{1}{2}$$

$$S_{\alpha_1 \alpha_4 \beta_4}^{Q} = -\frac{1}{2} , \ S_{\alpha_1 \beta_1}^{Q} = \frac{1}{2} , \ S_{C_1}^{Q} = \frac{1}{2} , \ S_{\alpha_2 \beta_2}^{Q} = 1$$

$$S_{R_2}^{Q} = 1 , \ S_{R_1 R_3 C_2}^{Q} = -\frac{1}{2}$$

Above result implies a good sensitivity performance of the proposed circuit, since all the active and passive sensitivity coefficients are obtained within less than unit.

3 Simulation results

The proposed mixed-mode universal filter presented in Fig. 3 is stimulated through TSPICE using the 0.18 μm TSMC CMOS technology process available from MOSIS at 25ºC. In simulation, the CCII is realized using the MOS implementation as shown in Fig. 2 with the dimensions transistors are taken as specified in Tab. 2.

This circuit is powered at low supply voltage ±0.75 V. It has a rail to rail dynamic range, a good accuracy, low resistor at borne X (9.14 Ω) and a wide bandwidth current mode (2.53 GHz) and voltage mode (3.53 GHz) [16].

Tab. 2. Aspect ratios of the transistors.

Transistors	$W\,(\mu m)$ / $L(\mu m)$
M_1, M_2, M_9	5/0.18
$M_3, M_4, M_5, M_6, M_7, M_8$	10/0.18
M_{10}	1/0.18
$M_{11}, M_{12}, M_{13}, M_{14}$	0.27/0.18
M_{15}	1.5/0.18

The proposed filter is designed with the passive element values, $R_1 = R_2 = R_3 = 1$ kΩ and $C_1 = C_2 = 5$pF. The voltage and current mode simulation results, high-pass, band-pass, all-pass, reject-pass, low-pass, of gain and phase responses have presented respectively in Fig. 4 and Fig. 5.

The simulated frequency responses have agreed well with the theoretical ones as expected, where as the difference between them arises from non-idealities such as non- ideal gain and parasitic impedance effects of the CCII.

The gain variation of voltage and current mode band-pass filters for different values of R_2 are presented respectively in Fig. 6.a and Fig. 6.b. It is shown that the quality factor can be adjusted by R_2, as depicted in equation (4) without affecting the pole frequency and with a variation of current gain as shown in (1).

The pole frequency variation of voltage and current mode band-pass filters for different values of capacitors ($C_1 = C_2$) are presented respectively in Fig. 7.a and Fig. 7.b. These result shows that the pole frequency can be adjusted without affecting the quality factor, as described in equation (3).

4 Conclusion

The multi-inputs single-output mixed mode multifunction filter based on four second generation current conveyor circuits, two grounded capacitors and four resistances is presented. The advantages of the proposed circuit are that: (i) it performs low-pass, high-pass, band-pass, band-pass, band-stop and all-pass functions dependent on an appropriate selection of three input signals of each mode, (ii) the natural angular frequency (ω_0) and the quality factor (Q) can also be controlled independently (iii) low active and passive sensitivity performances. The simulations results with TSPICE using the 0.18 μm TSMC CMOS technology have a good accurate with the theoretical.

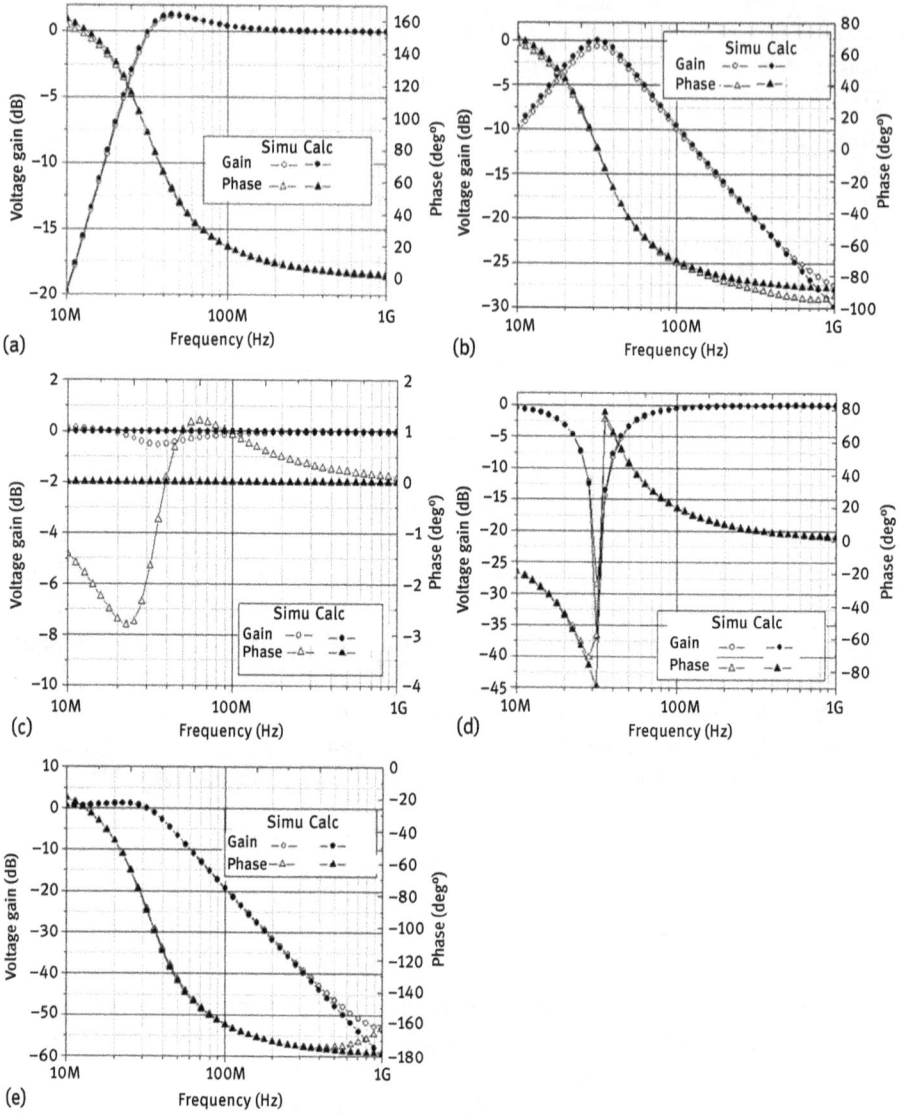

Fig. 4. Voltage mode simulation results of mixed mode universal filter: (a): High pass, (b): Band pass, (c): All pass, (d): Reject pass, (e): Low pass.

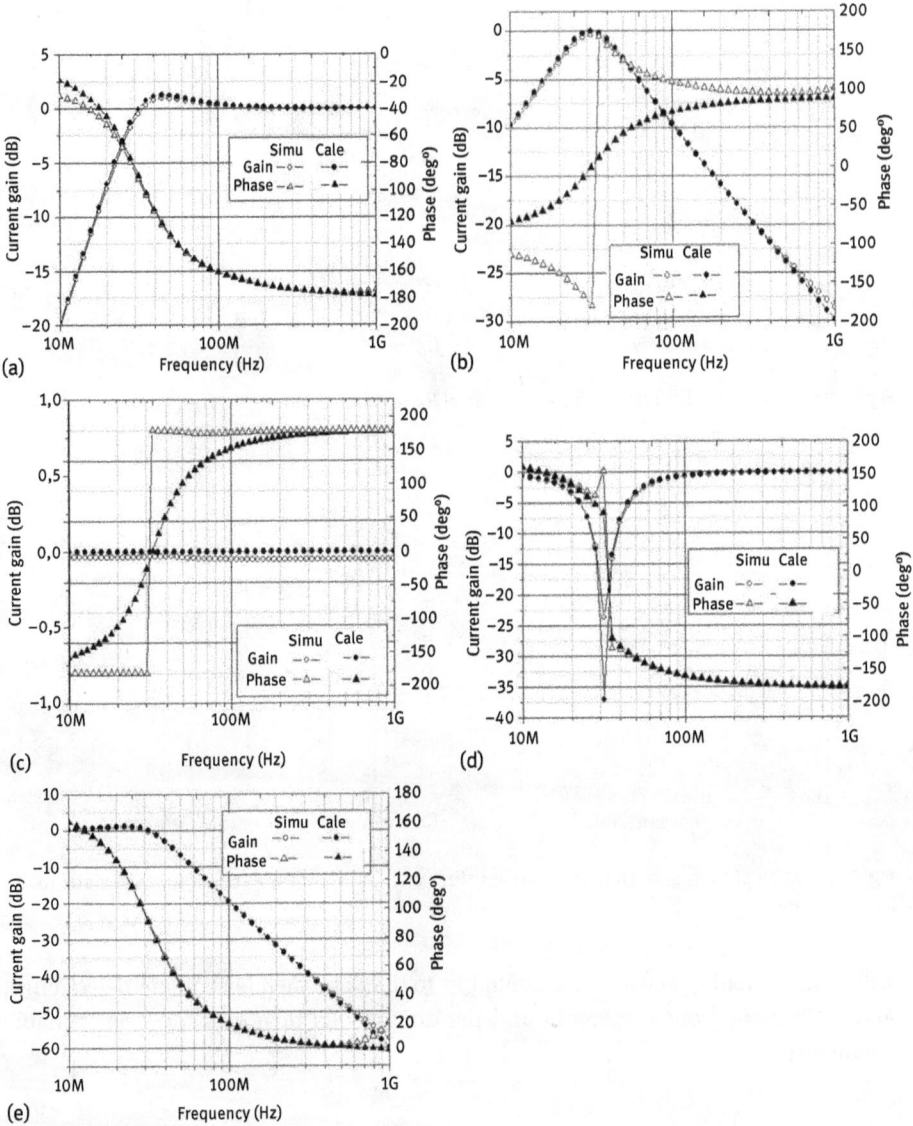

Fig. 5. Current mode simulation results of mixed mode universal filter: (a): High pass, (b): Band pass, (c): All pass, (d): Reject pass, (e): Low pass.

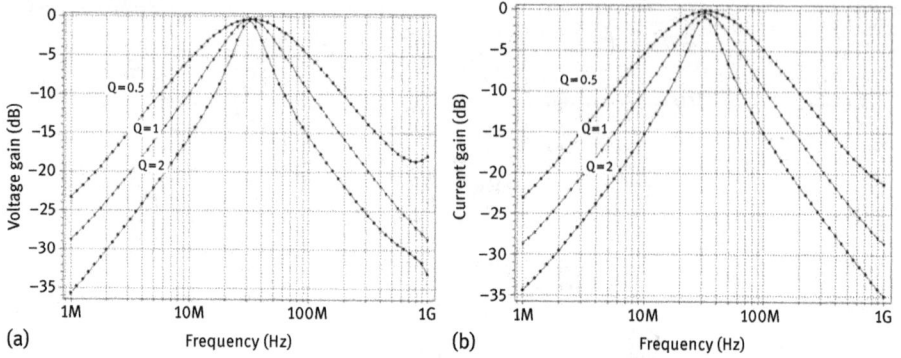

(a)

(b)

Fig. 6. Variation of quality factor different values of R2 (a) voltage mode (b) current mode.

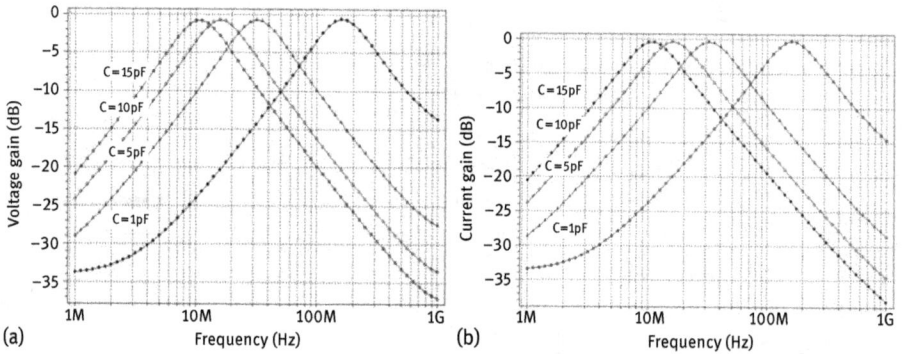

(a)

(b)

Fig. 7. Variation of frequency F0 with different values of capacitors ($C_1 = C_2$) (a) voltage mode (b) current mode.

Acknowledgments: The authors would like to thank all members of microelectronics and instrumentations laboratory and the anonymous reviewers for their valuable comments.

Bibliography

[1] Jiun-Wei Horng. High output impedance current-mode universal biquadratic filters with five inputs using multi-output CCIIs, *Microelectronics Journal*, 42 (5):693–700, 2011.

[2] M. Kumngern. Kobchai Dejhan. Electronically tunable voltage-mode universal filter with three-input single-output. *Int. conf. on Electronic Devices, Systems and Applications* (ICEDSA), :7–10, 2010.

[3] Jiun-Wei Horng. Current mode universal biquadratic filters with five inputs and one output using three ICCIIs. *Indian journal of pure and applied physics*, 49:214–217, 2011.

[4] Chun-Ming Chang. Universal voltage-mode filter with four inputs and one output using two CCIIs. *Inte. journal of electronics*, 83:305–309, 2013.

[5] S. Topaloglu. M. Sagbas. F. Anday. Three-input single-output second-order filters using current-feedback amplifiers. *Int. journal of electronics and communications*, 66:683–686. 2012.

[6] Chunhua Wanga. Jing Xua. Ali Ümit Keskinb. Sichun Dua. Qiujing Zhanga. A new current-mode current-controlled SIMO-type universal filter. *Int. Journal of Electronics and Communications* (AEÜ), 2010.

[7] Nejib Hassen. Thouraya Ettaghzouti. Kamel Besbes. High-performance second-generation controlled current conveyor CCCII and high frequency applications. *World Academy of Science Engineering and Technology*, 60:1361–1370. 2011.

[8] Hua-Pin Chen. Voltage-Mode Multifunction Biquadratic Filter with One Input and Six Outputs Using Two ICCIIs, *Scientific World Journal*. 2014..

[9] Malek Ramezani. Nima Ahmadpoor. New current-mode universal filter by a novel low voltage second generation current conveyor, *Microelectronics and Solid State Electronics*, 2(3)52–57. 2013.

[10] Wang Chunhua. Ali Umit Keskin. LengYang. Zhang Qiujing. Minimum configuration insensitive multifunctional current mode biquad using current conveyors and all grounded passive components. *Radioengineering*, 19, 2010.

[11] Jiun-Wei HORNG. To-Yao CHIU. Zih-Yang JHAO. Tunable versatile high Input impedance voltage-mode universal biquadratic filter based on DDCCs. *Radioengineering*, 21(4), 2012.

[12] Hua-Pin Chen. Sung-Shiou Shen, A Versatile Universal Capacitor-Grounded Voltage-Mode Filter Using DVCCs. *ETRI Journal*, 29(4), 2007.

[13] Muhammad Taher, A NOVEL mixed-mode current-controlled current- conveyor-based filter, *Active and Passive Electronic Components*, 26:185–191. 2003.

[14] Muhammad Taher Abuelma'atti. Abdulwahab Bentrice. A novel mixed-mode CCII based filter. *Active and passive electronic components*. 27:197–205, 2004.

[15] Neeta Pandey. Sajal K. Paul. Asok Bhattacharyya. S. B. Jain. A new mixed mode biquad using reduced number of active and passive elements. IEICE *Electronics Express*, 13(6):115–121, 2006,

[16] Thouraya Ettaghzouti. Nejib Hassen. Kamel Besbes. Novel second generation current conveyor and voltage mode universal filter application. 12th *Int. Multi-Conf. on Systems, Signals & Devices* (SSD), 2015.

Biographies

Thouraya Ettaghzouti was born in Tozeur, Tunisia, in 1983. She received the B.S. degree from the Faculty of Sciences of Monastir in 2008, the M.S. degree from at the same University at the Microelectronic and Instrumentation Laboratory in 2010. Actually, she is preparing the Ph.D degree. She is interested to the implementation of low voltage low power integrated circuit design.

Néjib Hassen was born in 1961 in Moknine, Tunisia. He received the B.S. degree in EEA from the University of Aix-Marseille I, France in1990, the M.S. degree in Electronics in 1991 and the Ph.D. degree in 1995 from the University Louis Pasteur of Strasbourg, France. From 1991 to 1996, he has worked as a researcher in CCD digital camera design. He implemented IRDS new technique radiuses CCD noise at CRN of GOA in Strasbourg. In 1995, he joined the Faculty of Sciences of Monastir as an In 1995, he joined the Faculty of Sciences of Monastir as an Assistant Professor of physics and electronics Since 1997, he has worked as researcher in mixed-signals neural networks. Currently, he is professor of microelectronics and electronics to ISIMM University of Monastir. He is focusing on the implementation low voltage - low power mixed and analog circuits.

Kamel besbes, Professor on Microeleronics, received M.S. degree from the Ecole Centrale de Lyon-France in 1986, the PhD degree from INSA Lyon, France in 1989 and the "State Doctorate Degree" from Tunis University in 1995. In 1989, he joined Monastir University. He established teaching and research laboratories initiatives in microelectronics since 1990. Research efforts are focused on microelectronics devices, microsystems, embedded systems, Instrumentation for detection, navigation and space programs. He has more than 200, published and presented papers at workshops and conferences. He participated to committees of several workshops and conferences. He is a full Professor since 2002 and he was the Vice-Dean (2000–2005), the Dean of Sciences Faculty of Monastir (2008–2011). He was elected member of University of Monastir council (2005–2014) and member of Higher Education and Scientific Research Reform National Council (2012–2014) and several national strategic committees and H2020 Space Tunisia-NCP. He is now the head of the Microelectronics and Instrumentation Lab(since 2003) in the University of Monastir and the General Director of the Centre for Research on Microelectronics and Nanotechnology in Sousse Technopark (since 2014).

T. Ettaghzouti, N. Hassen and K. Besbes

High Performance Second Generation Current Conveyor Circuit and Multiplier Applications

Abstract: In this paper, a high-performance CMOS second generation current conveyor circuit CCII operated with a low supply voltage (± 0.75 V) is presented. The input stage of this circuit is based on a super class AB CMOS OTA cells using an adaptive biasing circuit. All simulation results are performed by Tspice based on BSIM3v3 transistor model (level 49) for the TSMC 0.18 µm CMOS process available from MOSIS at 25°C. The proposed CCII shows a good linearity over the dynamic range with an excellent accuracy, low parasitic resistor at terminal X ($R_X = 8.26\Omega$), high input impedance at terminal Y and wide current mode (2.96 GHz) and voltage mode (3 GHz) bandwidth. Based on this circuit, a voltage and current mode bi-quadrant multiplier circuits are proposed. These applications have a large dynamic range and good agreement with the theoretical calculations.

Keywords: Second generation current conveyor CCII, accuracy, FVF, voltage mode, current mode, active multiplier.

1 Introduction

Analog multipliers are considered as basic building blocks in many systems and applications such as adaptive filters, fuzzy control, modulators, demodulators, analog signal processing and instrumentations. It can be divided into three groups depending on the operating mode (voltage, current and mixed mode). In the literature, many active multiplier configurations using a second generation current conveyor (CCII) have received a considerable attention [1, 2]. For this reason, several research works are carried to improve the CCII circuits. In the order to broaden the voltage mode dynamic range, the designers have used two pairs differential NMOS and PMOS connected in parallel [3, 4]. The operation of this stage can be divided into three regions. In the positive rail region, only NMOS pair is active. In the mid-rail region, both NMOS and PMOS are active. However in the negative rail region, only PMOS pair is active. Furthermore, Ahmed H. Madian and al. have used two NMOS pairs differential and two classic voltage inverter circuits in the input stage [5]. In which, the gates of two NMOS transistors, that are on the right, are linked respectively to the

T. Ettaghzouti, N. Hassen and K. Besbes: Micro-electronics and instrumentation laboratory University of Monastir Monastir, Tunisia, e-mails: thourayataghzouti@yahoo.fr, nejib.hassen@fsm.rnu.tn, kamel.besbes@fsm.rnu.tn.
K. Besbes: Centre for Research on Microelectronics and Nanotechnology of Sousse, Technopole of Sousse, Tunisia, e-mails: kamel.besbes@fsm.rnu.tn.

De Gruyter Oldenbourg, ASSD – Advances in Systems, Signals and Devices, Volume 6, 2018, pp. 65–80.
https://doi.org/10.1515/9783110448375-005

input and the output of the voltage inverter circuit, even case for the transistors on the left. In 2002, the researchers have discussed and developed many techniques in order to improve the bandwidth of CCII and to propose a new structure with greater bandwidth than traditional circuits [6].

In this paper, a high performance low-voltage (LV) low-power (LP) second current conveyor circuit CCII is presented. The input stage is based on a super class AB CMOS OTA cells. In the order to improve the accuracy of CCII an adaptive biasing has been used. This circuit is powered with a low supply voltage (± 0.75 V) and simulated by Tspice based on BSIM3v3 transistor model (level 49) for the TSMC 0.18 μm CMOS process available from MOSIS at 25°C. The proposed CCII presents a rail to rail dynamic range with a good accuracy. It has a low parasitic resistance at terminal X (8.26 Ω), high impedance at terminal Y and a wide current mode (2.96 GHz) and voltage mode (3 GHz) bandwidth.

Based on this circuit, a voltage and current active multipliers using a second generation current conveyor circuits and two NMOS transistors operating in linear region are presented. These applications have a large dynamic range and a good accuracy with the theoretic results.

2 Proposed second generation current conveyor circuit

2.1 Description circuit

Recent works are interested in improving the second generation current conveyor CCII circuit. A good circuit CCII has a good voltage and current mode linearity, excellent accuracy, wide bandwidth, low power consumption, low parasitic impedance at terminal X and high impedance at the terminal Y and Z.

The input block of CCII is the important wrench to obtain a large dynamic range. It is generally composed by trans-conductor circuit biased by a source degeneration using resistors or MOS transistors [7, 8], crossing-coupling of multiple differential pair [9], class AB configuration [10, 11], adaptive biasing [12, 13] and constant drain-source voltages [14].

The proposed second generation current conveyor using adaptive biasing is presented in Fig. 1.

The diagram adaptive biasing circuit is shown in Fig. 2a [15, 16]. The two transistors M_1 and M_2 are inter-connected by two DC voltages with the purpose of having a constant source-grid voltage (VSG1, 2).

Figure 2(b) shows an efficient implementation of very low impedance voltage source which offers more accurate shift of voltage level than a conventional source follower buffer. Each level shifter is built with "Flipped Voltage Follower" FVF cell.

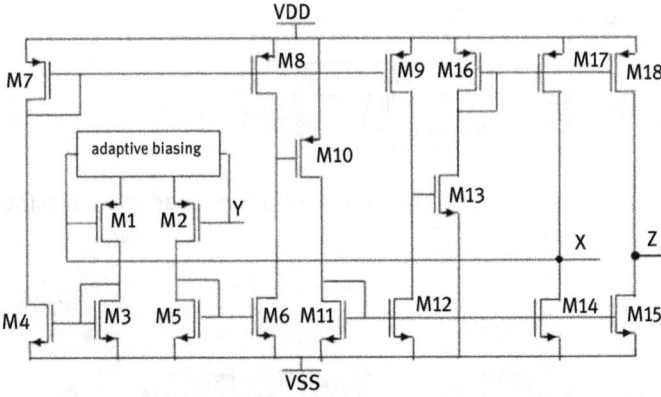

Fig. 1. Novel second generation current conveyor CCII.

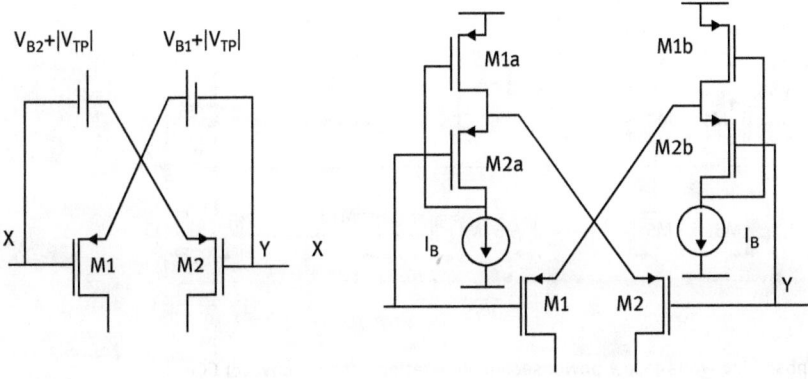

Fig. 2. Adaptive biasing circuit: (a) Diagram (b) Implementation using FVF.

FVF cells [17, 18] are composed of two transistors (M_{1a}, M_{2a} and M_{1b}, M_{2b}) and a current source IB.

The pair transistors M_1 and M_2 are operated in saturated region. The currents I_{D1} and I_{D2} are given by:

$$I_{D1,2} = \frac{\mu_P C_{Ox}}{2} \left(\frac{W}{L}\right)_{M_{1,2}} (V_{SG1,2} - |V_{TP}|)^2 \tag{1}$$

The relationship between V_{SG1} and V_{SG2} as a function of V_X and V_Y are given by:

$$V_{SG1} = (V_Y - V_X) + V_{B1} + |V_{TP}| \tag{2}$$

$$V_{SG2} = (V_X - V_Y) + V_{B2} + |V_{TP}| \tag{3}$$

Based on these two equations, voltages $V_{B1,2}$ are given by:

$$V_{B1,2} = \sqrt{\frac{2I_B}{\mu_P C_{Ox} \left(\frac{W}{L}\right)_{M_{2a,2b}}}} \tag{4}$$

The proposed second generation current conveyor circuit using an adaptive biasing circuit presented in Fig. 2b is shown in Fig. 3.

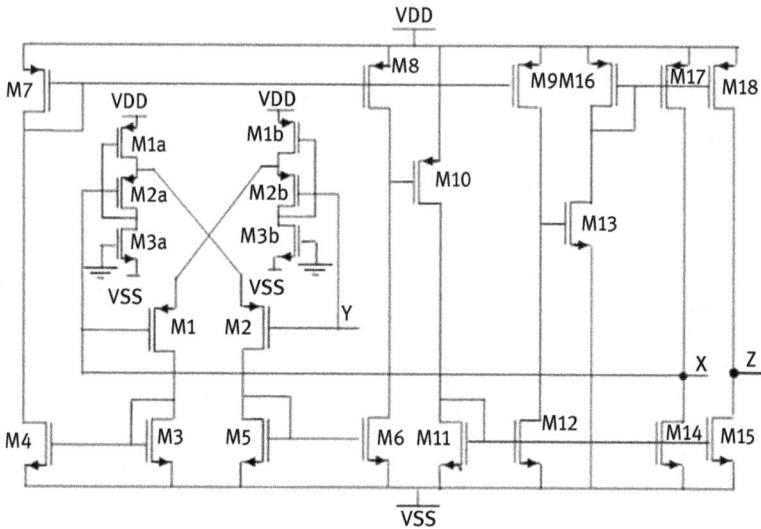

Fig. 3. Proposed low-voltage low-power second generation current conveyor CCII.

The input stage is composed by differential pair transistors (M_1, M_2) and three current mirrors formed by two matched transistors (M_3, M_4), (M_5, M_6) as well as (M_7, M_8).

To have a good voltage and current mode answers circuit, the dimensions of transistors have been estimated based on some conditions. Firstly, the drains of transistors M_7 and M_8 are respectively bound to the drain transistors of M_4 and M_6. This means that, the drain currents of M_1 and M_2 given in (5, 6) are equal.

$$I_{D1} = \frac{\mu_P C_{Ox}}{2} \left(\frac{W}{L}\right)_{M_1} \left(\sqrt{\frac{2I_B}{\mu_P C_{Ox} \left(\frac{W}{L}\right)_{M_{2b}}}} + V_Y - V_X\right)^2 \tag{5}$$

$$I_{D2} = \frac{\mu_P C_{Ox}}{2} \left(\frac{W}{L}\right)_{M_2} \left(\sqrt{\frac{2I_B}{\mu_P C_{Ox} \left(\frac{W}{L}\right)_{M_{2a}}}} - V_Y + V_X\right)^2 \tag{6}$$

Based on these two expressions, the difference potential between terminals X and Y is given by:

$$V_Y - V_X = \frac{\sqrt{\frac{2I_B}{\mu_P C_{Ox}}} \left[\sqrt{\left(\frac{L}{W}\right)_{M_{2a}}} - \sqrt{\left(\frac{W}{L}\right)_{M_1} \left(\frac{L}{W}\right)_{M_2} \left(\frac{L}{W}\right)_{M_{2b}}}\right]}{1 + \sqrt{\left(\frac{W}{L}\right)_{M_1} \left(\frac{L}{W}\right)_{M_2}}} \tag{7}$$

It is necessary that $\left(\frac{W}{L}\right)_{M_1} = \left(\frac{W}{L}\right)_{M_2}$ and $\left(\frac{W}{L}\right)_{M_{2q}} = \left(\frac{W}{L}\right)_{M_{2b}}$, so that the potential difference is equal to zero ($V_X = V_Y$). On the other side, the drains of M_{11} and M_{12} are connected respectively to the drains of M_{10} and M_9, which implies that the current $I_{D8} = I_{D10}$.

$$I_{D8} = \frac{\mu_P C_{Ox}}{2} \left(\frac{W}{L}\right)_{M_8} (V_{SG8} - |V_{TP}|)^2 \tag{8}$$

$$I_{D10} = \frac{\mu_P C_{Ox}}{2} \left(\frac{W}{L}\right)_{M_{10}} (V_{SG10} - |V_{TP}|)^2 \tag{9}$$

Since the equations (8) and (9) are equal, it is necessary that $V_{SG8} = V_{SG10}$ and $\left(\frac{W}{L}\right)_{M_8} = \left(\frac{W}{L}\right)_{M_{10}}$. The current mirror ($M_{11}$, M_{12}, M_{14}, M_{15}) and (M_{16}, M_{17}, M_{18}) have the same current because the drains of transistors M_{14} and M_{15} are connected respectively to the drains transistors M_{17} and M_{18}. This means that $I_{D13} = I_{D11}$.

$$I_{D13} = \frac{\mu_P C_{Ox}}{2} \left(\frac{W}{L}\right)_{M_{13}} (V_{SG8} - V_{TN})^2 \tag{10}$$

$$I_{D11} = \frac{\mu_P C_{Ox}}{2} \left(\frac{W}{L}\right)_{M_{11}} (V_{SG11} - V_{TN})^2 \tag{11}$$

The equality between these two currents (I_{D13}, I_{D11}) is approved only if $V_{GS11} = V_{GS13}$ and $\left(\frac{W}{L}\right)_{M_{11}} = \left(\frac{W}{L}\right)_{M_{13}}$. When all these conditions are confirmed, the current applied to X terminal will be copied in the Z terminal.

$$I_X = I_{D14} - I_{D17} = I_{D15} - I_{D18} = I_Z \tag{12}$$

2.1.1 Dynamic study of circuit

Figure 4 presents the small signal schema of proposed second generation current conveyor circuit.

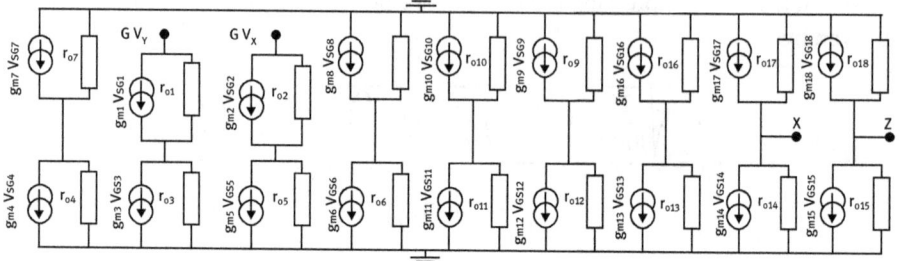

Fig. 4. Small signal circuit diagram presentation.

The input stage consists of an adaptive bias circuit based on flipped voltage follower FVF ($M_{1a,1b}, M_{2a,2b}, M_{3a,3b}$) while the current in the transistor $M_{3a,b}$ is kept constant. The gain G of FVF circuit is equal to:

$$G = \frac{V_{out}}{V_{in}} = \frac{(1 + g_{m1a,b}r_{o3a,b})g_{m2a,b}r_{o2a,b}r_{o1a,b}}{(1 + g_{m1a,b}r_{o3a,b})(1 + g_{m2a,b}r_{o2a,b}) + (r_{o2a,b} + r_{o3a,b})} \approx 1 \quad (13)$$

The transfer voltage gain A_v between nodes X and Y is given by:

$$A_v = \frac{G(1 + r_{o1}g_{m1})A_1 + r_{o2}g_{m2}A_2}{1 + r_{o1}g_{m1}A_1 + G(1 + r_{o2}g_{m2})A_2} \quad (14)$$

where:

$$A_1 = \frac{g_{m8}F_1R_1(g_{m14} + g_{m17}F_3R_2(1 + g_{m12}F_2R_3))}{F_4}$$

$$A_2 = \frac{R_1R_2R_3g_{m6}(g_{m14} + g_{m17}g_{m12}F_3F_2}{F_5}$$

and where:

$$R_1 = r_{o14}//r_{o17} \, , \quad R_2 = r_{o9}//r_{o12} \, , \quad R_3 = r_{o8}//r_{o6} \, , \quad F_1 = \frac{r_{o4}r_{o7}g_{m4}}{r_{o4} + r_{o7} + r_{o4}r_{o7}g_{m4}}$$

$$F_2 = \frac{r_{o11}r_{o10}g_{m10}}{r_{o11} + r_{o10} + r_{o11}r_{o10}g_{m11}} \, , \quad F_3 = \frac{r_{o13}r_{o16}g_{m13}}{r_{o13} + r_{o16} + r_{o13}r_{o16}g_{m16}}$$

$$F_4 = \frac{r_{o3} + r_{o1} + r_{o1}r_{o3}g_{m3}}{r_{o3}} \, , \quad F_5 = \frac{r_{o5} + r_{o2} + r_{o2}r_{o5}g_{m5}}{r_{o5}}$$

Using the approximation:

$$1 + r_{oi}g_{mi} \approx r_{oi}g_{mi}$$

we found: $F_1 = F_2 = F_3 = 1$, $F_4 = r_{o1}g_{m3}$, $F5 = r_{o2}g_{m5}$ and the gain bias G source becomes equal to unity. The transfer function has become:

$$A_v = \frac{Gg_{m1}g_{m8}R_1 \dfrac{g_{m14} + g_{m17}R_2(1 + g_{m12}R_3)}{g_{m3}} + g_{m2}R_1R_2R_3(g_{m14} + g_{m17}g_{m12})}{1 + g_{m1}g_{m8}R_1 \dfrac{g_{m14} + g_{m17}R_2(1 + g_{m12}R_3)}{g_{m3}} + Gg_{m2}R_1R_2R_3(g_{m14} + g_{m17}g_{m12})}$$

$$\approx 1 \tag{15}$$

Using this approximation, the parasitic resistance R_X is given by the following expression:

$$R_X = \frac{1}{R_3\left(g_{m14} + R_2g_{m12}g_{m17}\right)\left(g_{m2} + \dfrac{g_{m1}g_{m8}}{g_{m3}}\right) + \dfrac{g_{m1}g_{m9}g_{m17}R_2}{g_{m3}}} \tag{16}$$

2.1.2 Simulation results CCII

The performance of the proposed CMOS second-generation current conveyor was verified by Tspice based on BSIM3v3 transistor model (level 49) for the TSMC 0.18 µm CMOS process available from MOSIS at 25°C [19]. This circuit is powered by ±0.75 V and the transistors aspect ratios are given in Tab. 1.

The DC voltage transfer characteristic of CCII circuit is presented in Fig. 4. This circuit assured a good linearity over the rail to rail dynamic range (±0.75 V) with a maximum offset voltage equal to 0.012 mV. The frequency response between X and Y terminals is shown in Fig. 5. The CCII circuit has a unit voltage gain and a wide bandwidth of 3 GHz.

For testing the accuracy of CCII, firstly, a sinusoidal voltage with a variable amplitude and 100 MHz frequency has been applied at terminal Y. The value range of THD is variable from 0.003183 % to 0.01597 % (Fig. 6). Secondly, a sinusoidal voltage with a variable frequency and 0.35 V amplitude has been applied at terminal Y. The THD is less than a 0.11 % (Fig. 7).

Tab. 1. Aspect ratios of the transistors.

Transistors	W (µm) / L(µm)
M_1, M_2	5/0.18
$M_3, M_4, M_5, M_6, M_{11}, M_{12}, M_{13}, M_{14}, M_{15}$	1/0.18
$M_7, M_8, M_9, M_{10}, M_{16}, M_{17}, M_{18}, M_{1a}, M_{1b}$	2/0.18
M_{2a}, M_{2b}	3.5/0.18
M_{3a}, M_{3b}	0.27/0.18

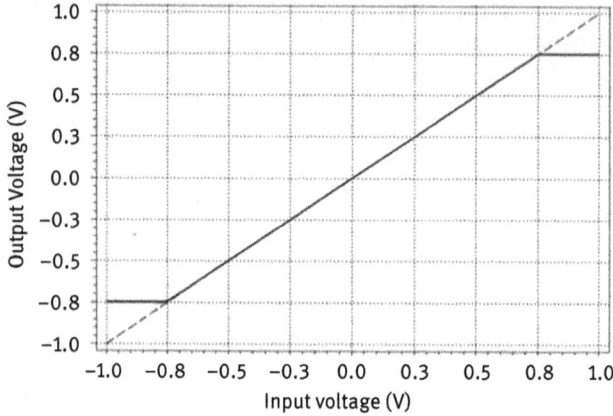

Fig. 5. Variation of output voltage as function of input voltage.

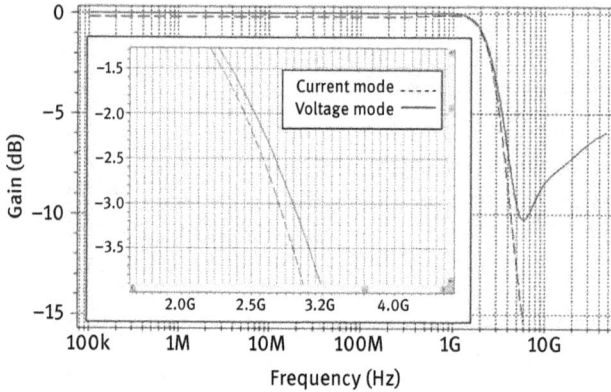

Fig. 6. Variation of voltage and current gain according to the frequency.

In current mode, a good linearity was obtained over the interval $[-220\,\mu A, 220\,\mu A]$ (Fig. 8). Fig. 5 shows a static gain of 0.99 and a cutoff frequency of 2.96 GHz. The parasitic elements on the tracks Y and Z are a resistor in parallel with a capacitor. They are given respectively $(R_Y//C_Y)$ ∞, 70 fF and $(R_Z//C_Z)$ 46.5 kΩ, 12.49 fF. On the other side, the input resistor at terminal X is equal to 8.26 Ω (Fig. 9).

The simulation results of proposed second generation current conveyor circuit with other reported in the literature have been grouped in table 2.

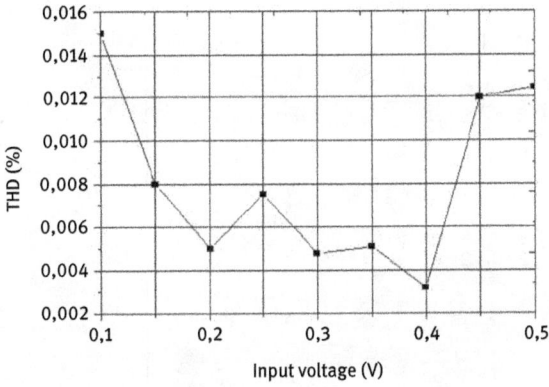

Fig. 7. THD variation as a function of the input voltage.

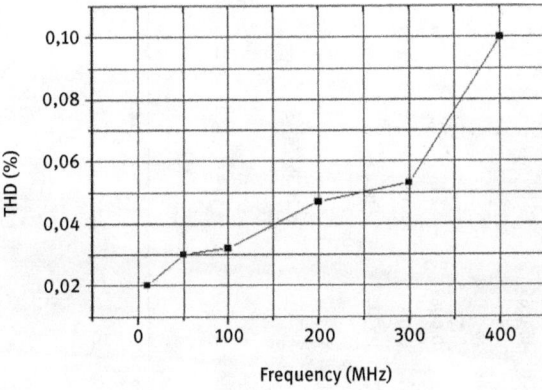

Fig. 8. THD variation as a function of the frequency.

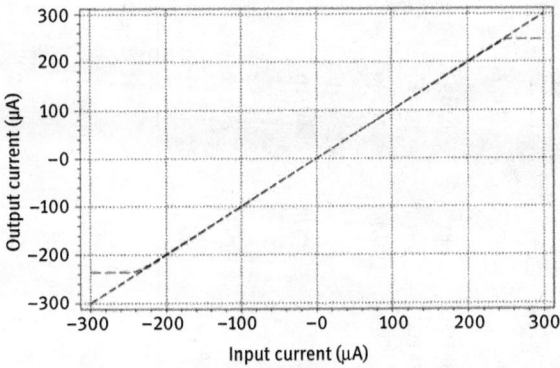

Fig. 9. Variation of output current as function of input current.

Tab. 2. The comparison table.

Characteristics	Unit	[20]	[21]	[22]	[23]	[24]	[25]	[26]	Proposed CCII
Technology	CMOS	0.18μm TSMC	0.35μm AMS	0.35μm AMS	0.18μm TSMC	0.18μm TSMC	0.35μm AMS	0.18μm TSMC	0.18μm TSMC
Bias voltage	V	±0.4	±0.75	±0.75	±1.25	±0.75	±1.65	±1	±0.75
Power consumption	mW	0.064	0.118	0.213	–	0.268	1.2	–	0.230
Voltage gain	–	1	1	–	0.96	1	1	0.948	1
Current gain	–	1	1	–	0.976	1	0.99	1	1
DC voltage range	V	-0.38 to 0.38	–	-0.65 to 0.65	–	-0.75 to 0.75	–	-0.4 to 0.4	-0.75 to 0.75
DC current range	μA	-7 to 7	–	–	–	-125 to 125	–	-350 to 350	-220 to 220
Bandwidth FCi	GHz	0.013	0.0105	0.0062	2.6	1.24	0.99	3.34	2.96
Bandwidth FCv	GHz	0.014	0.0105	0.0105	3.9	1.22	0.88	4.37	3
parasitic impedance R_X L_X	Ω//μH	27.860	13.04	<7, –	18.47	1.8	0.7, –	169.32	8.26//62.3
parasitic impedance R_Y C_Y	kΩ//fF	∞//1117	∞//500	–	25.35//49.5	–	∞, 62	5.67//164	Ω//12.49
parasitic impedance R_Z C_Z	kΩ//fF	890	2600/100	–	34.5/9.45	–	34.5,9.45	6.81//37.5	46.5//2.49

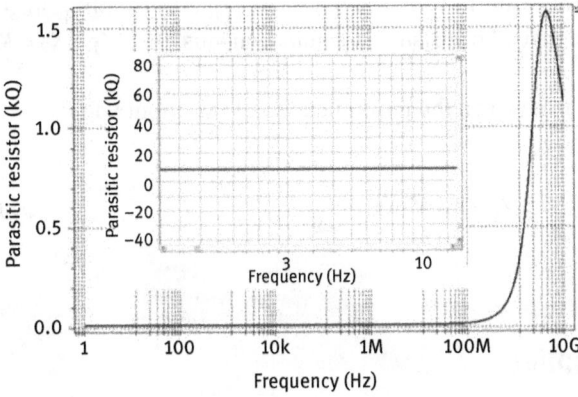

Fig. 10. Variation of parasitic resistance according to the frequency.

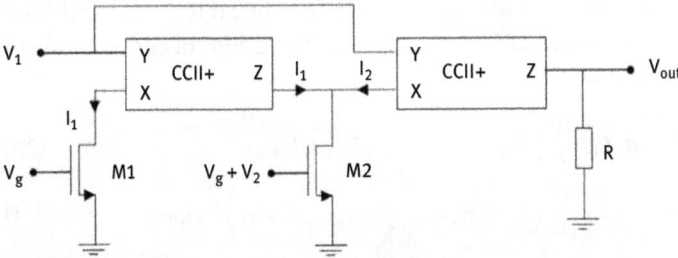

Fig. 11. Proposed voltage mode multiplier circuit.

3 Voltage and current multiplier circuits

3.1 Voltage mode multiplier

The proposed voltage mode multiplier circuit with two inputs and single output terminal is presented in Fig. 10. It is composed by two second generation current conveyor CCII circuits and two NMOS transistors.

Assuming that M_1 and M_2 are operated in liner region, the drain current expressions can be expressed as following:

$$I_1 = K_1 \left(V_{GS1} - V_{TN1} - \frac{1}{2} V_{DS1} \right) V_{DS1} \tag{17}$$

$$I_1 + I_2 = K_2 \left(V_{GS2} - V_{TN2} - \frac{1}{2} V_{DS2} \right) V_{DS2} \tag{18}$$

where: $K_1 = \mu_n C_{ox} \dfrac{W_i}{L_i}$ is the parameter of transistor, μ_n is the electron mobility, C_{ox} is the gate oxide capacitance per unit area, $\dfrac{W_i}{L_i}$ is the transistor aspect ratio, V_{GS} is the gate-to-source voltage, V_{DS} is the drain-to-source voltage and V_{TH} is threshold voltage of the MOS transistor. Suppose that all transistors are homogeneous, then $K_1 = K_2 = K$ and $V_{TH1} = V_{TH2} = V_{TH}$.

The output voltage (V_{out}) is given as the multiplication of two input voltages V_1 and V_2 with a gain G:

$$V_{out} = G V_2 V_1 \tag{19}$$

where: $G = R \mu_n C_{ox} \dfrac{W}{L}$.

3.2 Current mode multiplier

The proposed current mode multiplier circuit is composed by four second generation current conveyor circuits, two resistances and two NMOS transistors.

Assuring that the two NMOS transistors are operated in linear region and without considering the parasitic elements of CCII circuits, the expressions of current drain of M_1 and M_2 are given as following:

$$I_1 = \mu_n C_{ox} \frac{W}{L} \left(V_G - V_{TN} - \frac{1}{2} R_1 I_{in1} \right) R_1 I_{in1} \tag{20}$$

$$I_1 + I_{out} = \mu_n C_{ox} \frac{W}{L} \left(V_G + R_2 I_{in2} - V_{TN} - \frac{1}{2} R_1 I_{in1} \right) R_1 I_{in1} \tag{21}$$

Based on (20) and (21), the output current I_{out} is given by:

$$I_{out} = \mu_n C_{ox} \frac{W}{L} R_1 R_2 I_{in1} I_{in2} \tag{22}$$

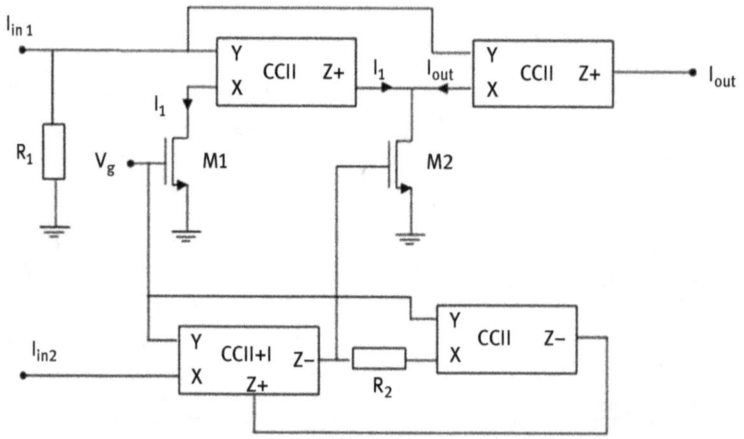

Fig. 12. Proposed current mode multiplier circuit.

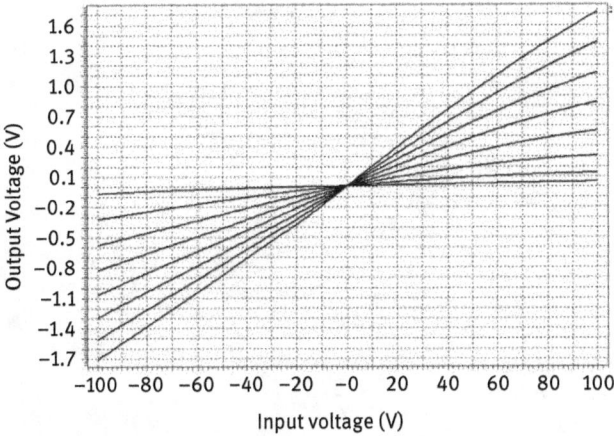

Fig. 13. DC transfer characterized of voltage mode multiplier.

3.3 Simulation results of multiplier circuits

The multiplier circuits presented in Fig. 10 and Fig. 11 are designed and simulated by using TSPICE using the 0.18 μm TSMC CMOS technology process available from MOSIS at 25°C.

To ensure that the transistors M_1 and M_2 are operated in linear region, the input voltage V_G is equal to 0.4 V and the length and the width of transistors are W = 50 μm, L = 5 μm, the resistances. By making the input voltage V2 Varied between 0 V and 0.35 V, the voltage mode multiplier has a good linearity over the range dynamic ±100 mV with a maximum offset voltage is equal to 0.02 mV, as shown in Fig. 12.

The simulation result of current mode multiplier is presented in Fig. 13. The dynamic range is extended between ± 10 μA with the offset current is less than 0.01 nA.

4 Conclusion

This paper presents a high performance CMOS second generation current circuit. In the input stage, the use of flipped voltage follower (FVF) circuit allowed us to obtain CCII circuit of a good linearity, high accuracy and low power consumption. The proposed CCII is operative at low supply voltage ±0.75 V with a reduced power consumption of 230 μW. It has a rail to rail dynamic range, a high accuracy, a low parasitic resistance of the terminal X (R_X = 8.26 Ω), a high input impedance at terminal Y and wide voltage mode (3 GHz) and current (2.96 GHz) bandwidth.

Based on this circuit, two voltage mode and current mode bi-quadratic multiplier circuits using a second generation current conveyor circuits and NMOS transistor

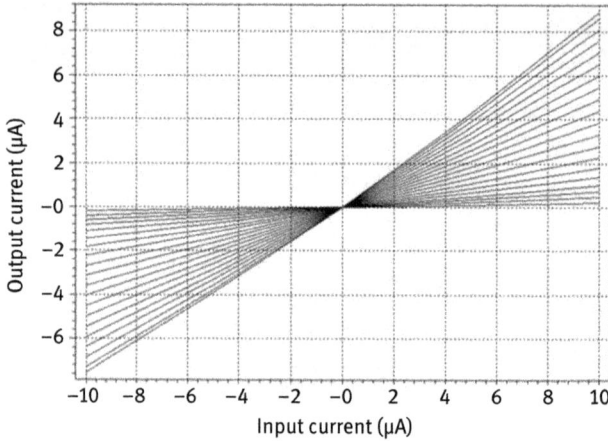

Fig. 14. DC transfer characterized of current mode multiplier.

operated in linear region are proposed. These two applications have a large dynamic range with a good accuracy.

Acknowledgments: The authors would like to thank all members of microelectronics and instrumentations laboratory and the anonymous reviewers for their valuable comments.

Bibliography

[1] Mohammad Biabanifard, S. Mehdi Largani, Ali Biabanifard and Javad Hosseini. Bulk-driven current conveyer based- CMOS analog multiplier. *Electrical and Electronics Engineering: An Int. Journal* (ELELIJ), 4(4):55–62, 2015.

[2] A. H. M. Abolila, H. F. A. Hamed and E. S. A. M. Hasaneen. New ±0.75 V low voltage low power CMOS current conveyor. *Microelectronics* (ICM), 2010.

[3] Emre ARSLAN. On the realization of high performance current conveyors and their applications. *Journal of Circuits, Systems, and Computers*, 2013.

[4] T. Ettaghzouti, N. Hassen and K. Besbes. Novel CMOS second generation current conveyor CCII with rail-to-rail input stage and filter application, *Multi-Conf. on Systems, Signals & Devices* (SSD), 2014.

[5] Ahmed H. Madian, Soliman A Mahmoud and Ahmed M. Soliman. New 1.5-V CMOS second generation current conveyor based on wide range transconductor, *Analog Integrated Circuits and Signal Processing*, 2006.

[6] Luís Nero Alves, Rui L. Aguiar and Dinis M. Santos. Bandwidth Aspects in Second Generation Current Conveyors, *Analog Integrated Circuits and Signal Processing*, (33):127–136, 2002.

[7] Josk Silva-Martinez, Michel S. J. Steyaert and Willy M. C. Sansen. A Large-Signal Very low-distortion transconductor for high-frequency continuous-time filters. *IEEE Journal Of Solid-State Circuits*, 27:1843–1853, 1992.

[8] Pietro Monsurrò, Salvatore Pennisi, Giuseppe Scotti and Alessandro Trifiletti. Linearization Technique for Source-Degenerated CMOS Differential Transconductors. *IEEE Trans. on Circuits and Systems*-II, 54:848–852, 2007.

[9] Chen J. Sanchez-Sinencio and E. Silva-Martinez J. Frequency-Dependent Harmonic-Distortion Analysis of a Linearized Cross-Coupled CMOS OTA and its Application to OTA-C Filters. *IEEE Trans. on Circuit and Systems*, 53:499–510, 2006.

[10] M. Laguna, C. De la Cruz-Blas, A. Torralba, R.G. Carvajal, A. Lopez- Martin and A. Carlosena. A novel low voltage low power class AB linear transconductor. *IEEE Int. Symp. on Circuits and Systems*, 1:725–728, 2004.

[11] J.A. Galan, R. G. Carvajal, F. Munog, A. Torralba and J. Ramirez-Angulo. Low power low voltage class AB linear OTA for HF filters with a large tuning range. *IEEE Trans. on Circuits and Systems*, 2:9–12, 2004.

[12] S. Sengupta. Adaptively biased linear transconductor. *IEEE Trans. on Circuits and Systems*, 52:2369–2375, 2005.

[13] M. G. Degrauwe, J.Rijmenants and E. A. Vittoz. Adaptive biasing CMOS amplifiers. *IEEE Journal of Solid State Circuits*, 17:522–528, 1982.

[14] Ayman A. Fayed and Mohammed Ismail. A Low-Voltage, Highly Linear Voltage-Controlled Transconductor. *IEEE Trans. on Circuits and Systems*-II. 52:831–835, 2005.

[15] Ramón González Carvajal, Jaime Ramírez-Angulo, Antonio J. Lopez-Martín and Antonio Torralba. The flipped voltage follower: a useful cell for low-voltage low-power circuit design. *IEEE Trans. on circuits and systems*-I, 52:1276–1291, 2005.

[16] Houda Bdiri Gabbouj, Néjib Hassen and Kamel Besbes. Low Voltage High Gain Linear Class AB CMOS OTA with DC Level Input Stage. *World Academy of Science, Engineering and Technology*, 56:8–26, 2011.

[17] Antonio J. Lopez-Martin and Alfonso Carlosena. Jaime Ramirez-Angulo. Very Low Voltage MOS Translinear Loops Based on Flipped Voltage Followers. *Analog Integrated Circuits and Signal Processing*, 40:71–74, 2004.

[18] J. Galan, R. G. Carvajal, A. Torralba, F. Muñoz and J. Ramirez-Angulo. A Low-Power Low-Voltage OTA-C Sinusoidal Oscillator with a Large Tuning Range. *IEEE Trans. on Circuits and Systems*-I, 52:283–291, 2005.

[19] Available from: http://www.mosis.org/Technical/Testdata/tsmc-018-prm.html

[20] Fabian Khateb, Nabhan Khatib and David Kubanek. Novel low-voltage low-power high-precision CCII based on bulk-driven folded cascade OTA. *Microelectronics Journal*, 42:622–631, 2011.

[21] Giuseppe Ferri, Claudia Di Carlo and Vincenzo Stornelli. A CCII-Based Low-Voltage low-power read-out circuit for dc-excited resistive gas sensors. *IEEE Sensors Journal*, 9:2035–2041, 2009.

[22] Ahmed H. Madian, Soliman A. Mahmoud and Ahmed M. Soliman. New 1.5-V CMOS second generation current conveyor based on wide range transconductor. *Analog Integrated Circuits and Signal Processing*, 49:267–279, 2006.

[23] Samir Ben Salem, Mourad Fakhfakh, Dorra Sellami Masmoudi Mourad Loulou Patrick Loumeau and Nouri Masmoudi. A high performances CMOS CCII and high frequency applications. *Analog Integrated Circuits and Signal Processing*, 49:71–78, 2006.

[24] H.M.A. Ahmed, H.F.A. Hesham and M.H. El-Sayed. High performance wideband CMOS current conveyor for low voltage low power applications. *Signal Processing & Information Technology*, 433–438, 2011.

[25] Emre Arslan, Shahram Minaei and Avni Morgul. On the realization of high performance current conveyors and their applications. *Journal of Circuits, Systems, and Computers*, 22:1–23, 2013.

[26] Néjib Hassen, Thouraya Ettaghzouti and Kamel Besbes. High-performance Second-Generation Controlled Current Conveyor CCCII and High Frequency Applications. *World Academy of Science, Engineering and Technology*, 60:12–24. 2011.

Biographies

Thouraya Ettaghzouti was born in Tozeur, Tunisia, in 1983. She received the B.S. degree from the Faculty of Sciences of Monastir in 2008, the M.S. degree from at the same University at the Microelectronic and Instrumentation Laboratory in 2010. Actually, she is preparing the Ph.D degree. She is interested to the implementation of low voltage low power integrated circuit design.

Néjib Hassen was born in 1961 in Moknine, Tunisia. He received the B.S. degree in EEA from the University of Aix-Marseille I, France in 1990, the M.S. degree in Electronics in 1991 and the Ph.D. degree in 1995 from the University Louis Pasteur of Strasbourg, France. From 1991 to 1996, he has worked as a researcher in CCD digital camera design. He implemented IRDS new technique radiuses CCD noise at CRN of GOA in Strasbourg. In 1995, he joined the Faculty of Sciences of Monastir as an In 1995, he joined the Faculty of Sciences of Monastir as an Assistant Professor of physics and electronics Since 1997, he has worked as researcher in mixed-signals neural networks. Currently, he is professor of microelectronics and electronics to ISIMM University of Monastir. He is focusing on the implementation low voltage - low power mixed and analog circuits.

Kamel besbes, Professor on Microeleronics, received M.S. degree from the Ecole Centrale de Lyon-France in 1986, the PhD degree from INSA Lyon, France in 1989 and the "State Doctorate Degree" from Tunis University in 1995. In 1989, he joined Monastir University. He established teaching and research laboratories initiatives in microelectronics since 1990. Research efforts are focused on microelectronics devices, microsystems, embedded systems, Instrumentation for detection, navigation and space programs. He has more than 200, published and presented papers at workshops and conferences. He participated to committees of several workshops and conferences. He is a full Professor since 2002 and he was the Vice-Dean (2000–2005), the Dean of Sciences Faculty of Monastir (2008–2011). He was elected member of University of Monastir council (2005–2014) and member of Higher Education and Scientific Research Reform National Council (2012–2014) and several national strategic committees and H2020 Space Tunisia-NCP. He is now the head of the Microelectronics and Instrumentation Lab(since 2003) in the University of Monastir and the General Director of the Centre for Research on Microelectronics and Nanotechnology in Sousse Technopark (since 2014).

W. Makni, M. Najari and H. Samet
Schottky Barrier Carbon Nanotube Transistors Op-Amp Circuit

Abstract: This paper presents a computationally efficient Raychowdhury compact model for the Schottky barrier (SB) carbon nanotube field-effect transistor (CNTFET). In order to achieve an accurate compact model, shot noise sources are added. Then, for the assessment of the SB on circuit performances, an operational amplifier (Op-Amp) has been designed using the SB-CNTFET compact model, and results have been compared with a conventional CNTFET.Noise analysis of the Op-Amp has been studied to determine the output noise. Choosing a suitable dielectric gate insulator material is important to optimize the output noise. An interesting case has been found when using oxide aluminum gate insulator with small thickness. An optimum noise performance has been achieved in this design.

Keywords: Carbon nanotube, field-effect transistor (CNTFET), Op Amp, Verilog-A, Compact model.

1 Introduction

Recently, the interest in novel materials able to overcome the miniaturization limits imposed by silicon based transistors has led researchers to explore alternative technologies such as carbon nanotubes. Carbon nanotubes are a very promising material for future electronics applications, both as interconnects and as field-effect transistors thanks to their low dimensionality, high mobility, ballistic transport and low power dissipation [1]-[2]. Although different types of CNTFETs have been fabricated and studied, the most important distinction between conventional CNTFET (C-CNTFET) and Schottky-barrier-type (SB-CNTFET) [3]-[4] is that C-CNTFET devices are characterized by doped CNT channels and Ohmic contacts, and this require more engineering challenges. However, SB-CNTFETs are easier to fabricate, since they use intrinsic CNTs channels with metallic drain and source contacts [5]. Schottky contact CNTFETs operate by the gate that modulates the transmission coefficient of Schottky barriers at the contact between the metal and the CNT. It is the quantum tunneling current through the SB that dominates over the thermionic current [6]. As a consequence, tunneling current causes shot noise which limits the electrical circuit

W. Makni, M. Najari and H. Samet: LETI Laboratory, National School of Engineering of Sfax, Sfax, Tunisia, e-mails: wafa_benayed@yahoo.fr, montassar.najari@gmail.com, hekmet.samet@enis.rnu.tn.
M. Najari: IKCE Unit, Jazan University, Jazan, Saudi Arabia, Department of Physics, Faculty of Sciences of Gabes, Gabes, Tunisia, e-mails: montassar.najari@gmail.com.

De Gruyter Oldenbourg, ASSD – Advances in Systems, Signals and Devices, Volume 6, 2018, pp. 81–92.
https://doi.org/10.1515/9783110448375-006

performances [7]. A computationally efficient compact model for the SB-CNFET is necessary for designing a large-scale circuit and for estimating and optimizing the circuit performance. In this context, this paper presents the implementation of an efficient Raychowdhury compact model [8] for the SB-CNTFET with the addition of shot noise source. An operational amplifier (Op-Amp) is designed using the SB-CNTFET compact model for the assessment of the SB impact on the circuit performances. Results are compared with a conventional CNTFET. Finally, noise analysis of the Op-Amp is examined to determine the output noise. An optimum output noise performance is achieved by using a suitable dielectric gate insulator material. Verilog-A language [9] is used to implement the CNTFET behavioral in the Advanced Design System environment [10].

This paper is organized as follows. Section II gives a brief description of the SB-CNTFET compact model. Section III shows the introduction of Shot noise in the SB-CNTFET compact model. Section IV investigates on analog circuit design application of an Op-Amp by pointing out the impact of the SB on circuit performances. Then, results are compared with Op-Amp using C-CNTFET. Section V presents a noise analysis of Op Amp and proposes a solution to minimize noise of the Op Amp with SB-CNTFET model. Finally Section V gives conclusions and perspectives of this work.

2 SB-CNTFET compact model

Contrary to the conventional CNFET, the SB-CNTFET operates not only by varying the channel potential but also by varying the source and drain transparency. This variable contact resistance is caused by the presence of two SBs at the source and drain metal/intrinsic CNT junctions that depend on the metal contact type (i. e., Pd, Ti, or Al), the CNT diameter, and the electrostatic environment. Moreover, the dimensional (height and width) characteristics of these SBs are modulated by the external biases applied to the device.

In SB-CNTFETs, depending on the work function difference between the metal contact and the CNT, carriers at the metal/CNT interface encounter different barrier heights. Therefore, carriers with energies above the SB height reach the channel by thermionic emission. On the other hand, carriers with energies below the SB height have a tunnel probability to reach the channel according to a transmission function [11].

The drain current is given by equation (1)

$$I_{DS} = \frac{4e}{h} \sum_{p=1}^{n_{bsbbd}} \left[\int_0^\infty f_S(E)T_S(E)dE - \int_0^\infty f_D(E)T_D(E)dE \right] \tag{1}$$

Where $f_S(E)$ and $f_D(E)$ are the Fermi-Dirac distribution function source and drain, respectively. The transmission coefficient of the SB at the source/CNT Interface $T_S(E)$ and CNT/drain $T_D(E)$ as a function of carrier energy is depicted from the Wentzel-Karmers-Brillouin (WKB) approximation as follows [12]:

$$T_S(E) = \exp\left[-A(\Phi_{SB} - E - qV_S)^{\frac{3}{2}}\right] \tag{2}$$

$$T_D(E) = \exp\left[-A(\Phi_{SB} - E - qV_{DS})^{\frac{3}{2}}\right] \tag{3}$$

where:

$$A = 4\frac{\sqrt{2m^*}}{3hqE_{elec}} \tag{4}$$

Φ_{SB} is the SB height, $E_{elec} = \dfrac{\Phi_{SB}}{t_{ox}}$ is the electrical field in the tunnel junction, t_{ox} is the oxide thickness and m^* is the carrier effective mass. Like in a Fabry-Pérot cavity, the total transmission function $T_T(E)$ over the whole structure is predicted as follows [13]:

$$T_T(E) = \frac{T_S(E)T_D(E)}{T_S(E) + T_D(E) - T_S(E)T_D(E)} \tag{5}$$

To overcome the complexity of implementing the draincurrent given by equation (1) in a compact model, an approximation method based on recent works from [14] isapplied. This approximation consists of analytically solvingthe 1-D modified Poisson equation for the channel potentialand, therefore, calculating an effective SB height. This effective SB height [14] is expressed as:

$$\Phi_{SB}^{S,D_p} = [\Phi_{SB} - (\Delta_p - eV_{CNT} + eV_{SD})]\exp\left(f() - \frac{d_{tunnel}}{\lambda_{Schottky}}\right) + (\Delta_p - eV_{CNT} + eV_{SD}) \tag{6}$$

where Δ_p: the equilibrium conduction band minimum of the pth sub-band is given by:

$$\Delta_p = \Delta_1 \frac{6p - 3 - (-1)^p}{4} \tag{7}$$

Δ_1 is the conduction band minimum for the first sub-band is set to half the Nanotube band gap $\Delta_p = 0.45/d$ (in eV).

V_{CNT} canal potential, d_{tunnel} is the tunneling distance [15], and $\lambda_{Schottky}$ is the screening length [16]. The screening length reflects the device geometry and in the case of a planar gate, is denoted by equation 8. Where ε_{nt} and ε_{OX} are respectively the CNT and oxide dielectric permittivity:

$$\lambda_{Schottky} = \sqrt{\frac{\varepsilon_{nt}}{\varepsilon_{OX}}\, d_{nt}\, t_{ox}} \tag{8}$$

The tunneling probability through the SB is set to unity if the barrier at some energy is thinner than d_{tunnel}, and is set to zero otherwise.

The current is calculated by means of the Landauer-Buttiker formula [17], assuming a 1-D ballistic channel in-between the SBs. Hence, after integration over energy and for all the energy sub-bands, the drain/source current is expressed as [8]. The tunnel current is the sum of all sub-band currents in conduction sub-bands. It can be calculated by the following expression:

$$I_{DS-SC} = \frac{2qM}{h} \sum_{p=1}^{n_{b-bands}} \left[\int_{\Phi_{SB-Sp}^{eff}}^{\infty} f_S(E)dE - \int_{\Phi_{SB-Dp}^{eff}}^{\infty} f_D(E)dE \right] \tag{9}$$

After integration, the drain current of Schottky barrier become

$$I_{DS-SC} = \frac{4K_BqT}{h} \sum_{p=1}^{\infty} \left[\ln \frac{1 + \exp \dfrac{qV_S - \Phi_{SB-Sp}^{eff}}{K_BT}}{1 + \exp \dfrac{qV_{DS} - \Phi_{SB-Dp}^{eff}}{K_BT}} \right] \tag{10}$$

SB-CNTFETs have an ambipolar characteristic as they conduct both electrons and holes, showing a superposition of $n-$ and $p-$ type behavior. The behavior is manifested in figure 1.

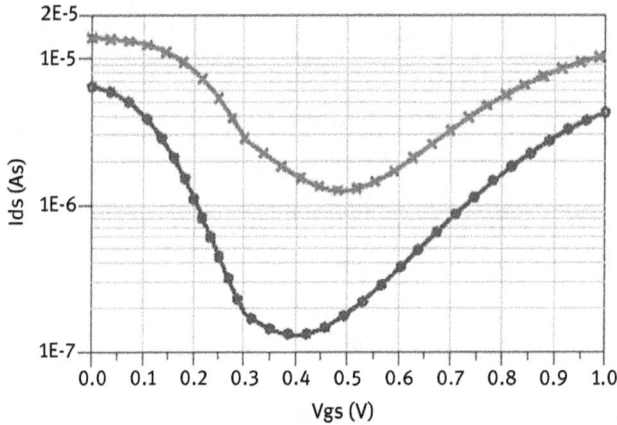

Fig. 1. I_{DS} versus V_{GS} for $V_{DS} = 0.1$ V (solid line with circle) and $V_{DS} = 0.2$ V (solid line with x) for ambipolar characteristics. With CNT diameter 1.48 nm and $\Phi_{SB} = 0.2$ eV.

For the case unipolar $N-$CNTFET or $P-$CNTFET the characteristic I_{DS} versus V_{GS} for $V_{DS} = 0.1$ V is given in Fig. 2 and Fig. 3, respectively.

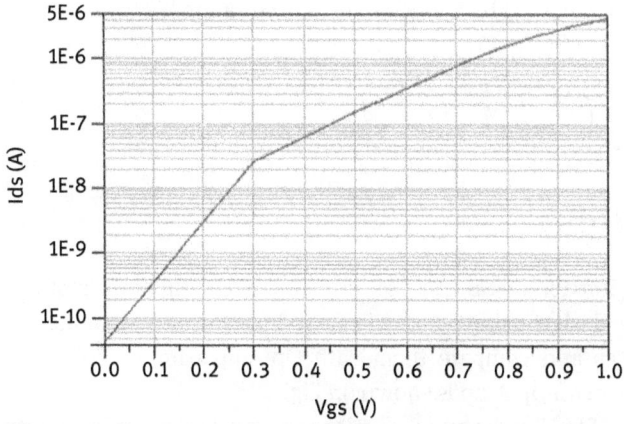

Fig. 2. I_{DS} versus V_{GS} for $V_{DS} = 0.1$ V graph for N-channel transistor with CNT diameter 1.48 nm and $\Phi_{SB} = 0.2$ eV.

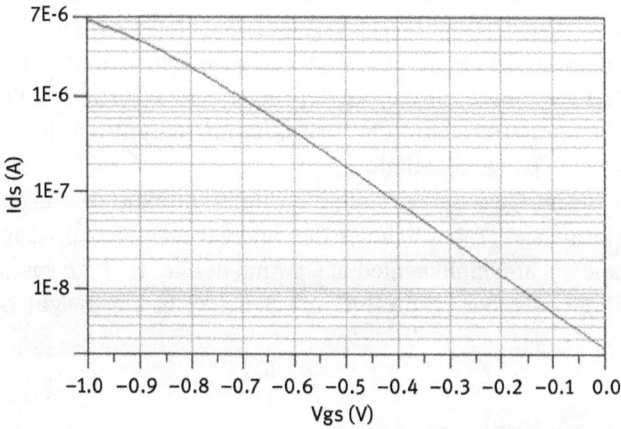

Fig. 3. I_{DS} versus V_{GS} for $V_{DS} = -0.1$ V graph for N-channel transistor with CNT diameter 1.48 nm and $\Phi_{SB} = 0.2$ eV.

3 Noise in SB-CNTFET

In order to obtain an accurate model, a noise analysis should be studied by adding convenient noise sources. Shot noise is the major noise source of SB-CNTFET transistor at high frequency. It is generated when carriers flow across a potential barrier of transistor [18]-[19]. This noise limits the performances of the system. Thermal noise is negligible compared to shot noise across the Schottky barrier. The power spectral density of this kind of noise, denoted by Sf is given by this equation [7]:

$$S_I(f) = \frac{4e^2}{h}\left[T_T(1 - T_T)eV_{DS}\coth\left(\frac{eV_DS}{2K_BT}\right)\right] \tag{11}$$

4 Circuit application

In order to analyze the impact of SB-CNTFET on analog circuit performances, two-stage Op-Amp circuit design using SB-CNTFET model has been simulated to determine its performances and compare them with Op-Amp using a C-CNTFET model.The main schematic of an Op-Amp is shown on Fig. 4.

The design of SB-CNTFET Op-Amp is tuned in order to have a compromise between the gain, unity gain bandwidth and power dissipation for high frequency application. Practical robust design values of load (CL) and miller compensation (Cc) capacitances are chosen as 0.1 fF and 0.01 fF, respectively.

The chosen values of these capacitances allow the stability in term of phase margin more than 45 degrees. The different CNTFET Op-Amp simulated parameters, using ADS environment are gain, phase margin , −3dB gain frequency, unity gain frequency, common mode rejection ratio (CMRR), Power Supply Rejection Ratio (PSRR), Input common mode voltage range (ICMR), V_{out} swing, output resistance (R_{out}) slew rate settling time and power dissipation.

Actions on the CNT diameter (d) were taken since diameter is the main parameter that affects SB-CNFET performances. Table 1 shows the different diameters used in Op-Amp design. These diameters are implemented in Op-Amp design, for both cases C-CNTFET and SB-CNTFET models. For SB-CNTFET the Schottky barrier height is $\Phi_{SB} = 0.01$ eV.

Tab. 1. Set of CNT's diameters using Op- Amp design.

Transistor	Diameter	(n,m)
$T_1 - T_2 - T_7$	1.48 nm	(19,0)
$T_3 - T_4 - T_6$	1.95 nm	(25,0)
$T_5 - T_8$	1.56 nm	(20,0)

Figure 5 Shows the results of gain and margin phase simulation for Op-Amp using SB-CNTFET. The gain is equal to 25.33 dB and it starts to cut off around 873 MHz (the −3 dB frequency). The gain bandwidth is equal to 16.81 GHz (the unity gain frequency, 0 dB).The phase margin is equal to 66°.

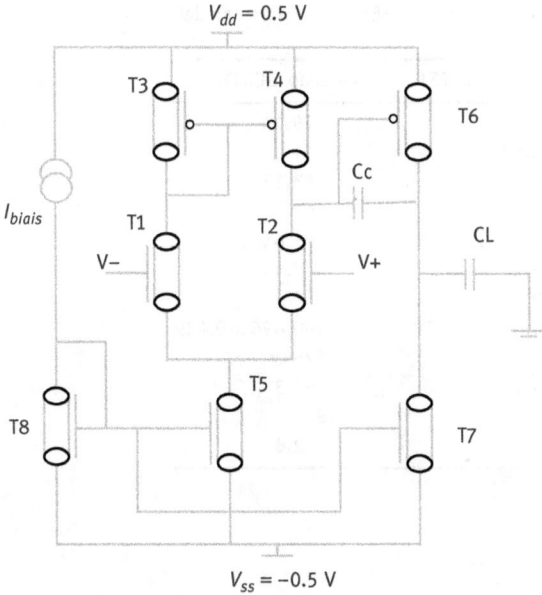

Fig. 4. Schematic of a basic two stage Op-Amp.

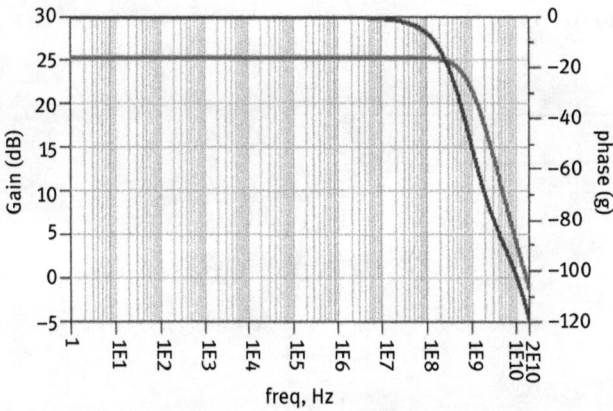

Fig. 5. (Right axis) gain, (left axis) Phase margin of SB-CNTFET Op-Amp.

Table 2 summarizes the comparison of the main characteristics of the Op-Amp performances with both SB-CNTFETand C-CNTFET models.Results show that by using SB-CNTFET, both gain and band widths of Op-Amp are decreased in comparison with C-CNTFET.

Tab. 2. Comparison of Op-Amp-performances for SB-CNTFET and C-CNTFET models.

Parameter	Op Amp SB-CNTFET	Op Amp C-CNTFET
Gain(dB)	25.33	29.14
−3dB frequency (GHz)	0.873	1
Unity gain frequency (GHz)	16.81	43.13
Phase margin (°)	66	67
ICMR (mV)	−281.8 to 396.2	−281.6 to 385.5
CMRR (dB)	49.44	49.44
PSRR (dB)	50	50
V_{out} swing (V)	−0.496 to 0.378	−0.496 to 0.429
R_{out} (kΩ)	125	125
Settling time (ps)	85.3	83.3
slew rate (V/μs)	9202	12896
Power dissipation (μW)	2.8	2.8

5 Noise analysis

In order to analyze the noise in the Op-Amp, the output noise was simulated by using a 10 nm thick of silicon dioxide (SiO2) with dielectric constant 3.9 used in the gate insulator. The Schottky barrier height is $\Phi_{SB} = 0.01$ eV. Figure 6 gives the output noise of Op-Amp circuit equals to 104.3 μV/$\sqrt{H}$.

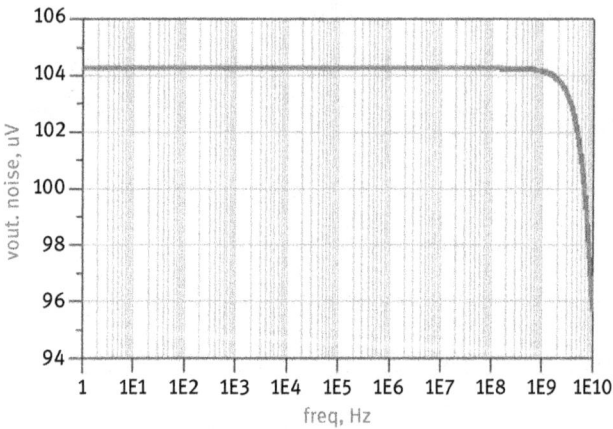

Fig. 6. Output noise of Op-Amp using 10 nm SiO2 gate insulator.

Increasing the (SiO_2) gate insulator oxide thickness allows for minimizing the output noise of Op-Amp. The simulation results are shown in Fig. 7 for different oxide thickness 10 nm, 15 nm and 20 nm.

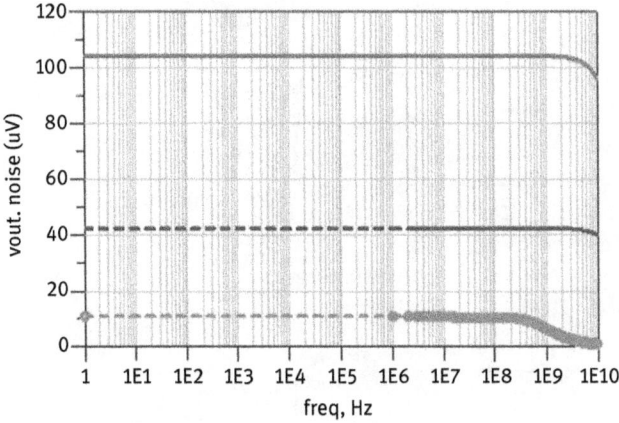

Fig. 7. Output noise of Op-Amp using SiO_2 gate insulator with different thickness 10 nm (red curve solid line), 15 nm (blue curve dashed line) and 20 nm (grey curve dashed line with circle).

Table 3 gives the output noise of Op-Amp using SiO_2 gate insulator 10 nm, 15 nm and 20 nm oxide thickness.

Tab. 3. Output noise of Op-Amp using SiO2 gate insulator with 10 nm, 15 nm and 20 nm oxide thickness.

SiO_2 thickness	Output noise ($\mu V/\sqrt[8]{H}$)
10 nm	104.3
15 nm	42.35
20 nm	10.64

As a second step, our purpose is to minimize the output noise. A higher dielectric constant material such as dioxide Aluminum (Al_2O_3) is used. This material has a dielectric constant (?=9) [20] and is used for the simulation of the output noise for different thickness (10 nm, 15 nm, 20 nm) as shown in figure 8 and 9. The values of the output noise for 10 nm, 15 nm and 20 nm oxide thickness are summarized in table 4.

Simulation results show that when the oxide thickness of (Al_2O_3) is small, the output noise decreases.

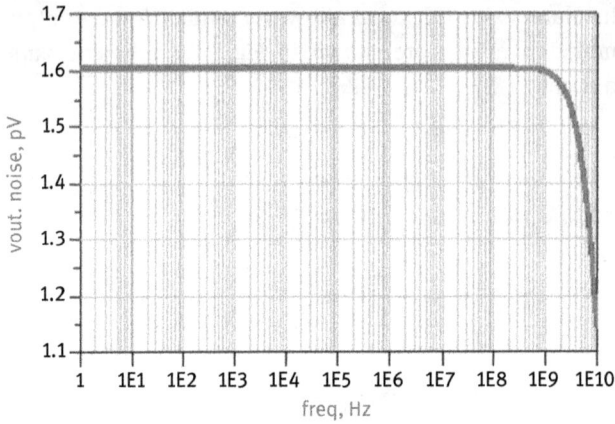

Fig. 8. Output noise of Op-Amp using Al2O3 for 10 nm oxide thickness.

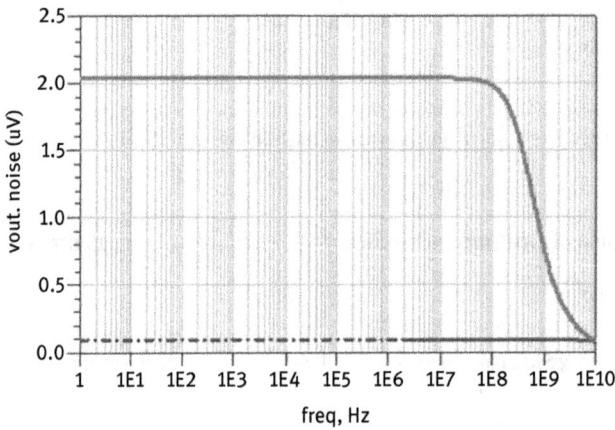

Fig. 9. Output noise of Op Amp using Al2O3 for different oxide thickness 15 nm (blue curve dashed line) and 20 nm (red curve solid line).

Tab. 4. Values of the output noise for different gate insulator materials, 10 nm, 15 nm and 20 nm oxide thickness.

Al$_2$O$_3$ thickness	Output noise
10 nm	1.6 pV/$\sqrt{H}$
15 nm	93.4 nV/$\sqrt{H}$
20 nm	2 μV/$\sqrt{H}$

6 Conclusion

In this paper, the case of Schottky barrier CNTFET compact model of Raychowdhry has been studied and the accuracy of the model has been improved by adding the shot noise. Operational Amplifier has been designed to determine the impact of SB on circuit performances and results are compared with a conventional CNTFET. Noise analysis of the Op-Amp has been investigated to determine the output noise. Results show that the use of a higher dielectric constant material such as dioxide Aluminum for gate insulator with small thickness is the best solution to minimize the output noise.

Bibliography

[1]   A. Javey, J. Guo, Q. Wang, M. Lundstrom and H. Dai. Ballistic carbonnanotube field-effect transistors. *Nature*, 424(6949), 2003.
[2]   W. Makni, M. Najari, H. Samet and M. Masmoudi. Operational amplifier circuit design using carbon nanotube transistors. *Nanoscience & Nanotechnology-Asia*, 3:106–113, 2013.
[3]   S. Heinze, J. Tersoff, R. Martel, V. Derycke, J. Appenzeller and Ph. Avouris. Carbon nanotubes as schottky barrier transistors. *Physics Review Letters*, 89(10), 2002.
[4]   C. Chen, D. Xu, E. Kong and Y. Zhang. Ballistic carbon nanotube field-effect transistors. *Nature*, 424:654–657, 2003.
[5]   S. Lee, M. Jang, Y. Kim, M. Jeon and K. Park.Schottky barrier Metal-Oxide-Semiconductor field-efffect transistors fornano-regime Applications. *Electronic Materials Letters*, 1(1):27–29, 2005.
[6]   P.Michetti and G.Iannaccone. Analytical model of one-dimensional carbon-basedschottky-barrier transistors. *IEEE Trans. Electron Devices*, 57(7):1616–1625, 2004.
[7]   M. J. M. De Jong and C. W. J. Beenakker. *Shotnoisein mesoscopic systems*. Kluwer Academic Publishers, 1997.
[8]   A. Raychowdhury, S. Mukhopadhyay and K. Roy. A circuit-compatiblemodel of ballistic carbon nanotube field-effect transistors. *IEEE Trans. Computer-Aided Design Integrated. Circuit Systems*, 23(10):1411–1420, 2004.
[9]   *Verilog-A Reference Manual*, Agilent Technologies September 2004.
[10] *Advanced Design System* (ADS) The website of ADS, http://www.home.agilent.com
[11] J. Appenzeller, M. Radosavljevi˜c, J. Knoch and P. Avouris. Tunnelingversus thermionic emission in one-dimensional semiconductors. *Physics Review Letters*, 92(4), 2004.
[12] W. C. Elmore and M. A. Heald. *Physics of Waves*. New York: Dover, 1985.
[13] S. Datta. *Electronic Transport in Mesoscopic Systems*. Cambridge, U.K.: Cambridge Univ. Press, 1997.
[14] J. Knoch and J. Appenzeller. Tunneling phenomena in carbon nanotube field-effect transistors. *Phys. Stat. Sol.* 205:679–694, 2008.
[15] J. Knoch and J. Appenzeller. Carbon nanotube field-effect transistors - The importance of being small. inAmIware Hardware Technology Drivers of Ambient Intelligence. Netherlands Springer 2006.

[16] R. Yan, A. Ourmazd and K. Lee. Scaling the Si MOSFET: From bulk to SOI to bulk. *IEEE Trans. Electron Devices*, 39(7):1704–1710, 1992.
[17] M. Anantram, M. Lundstrom and D. Nikonov. Modeling of nanoscale devices. *Proceedings of IEEE*, 96(9):1511–1550, September. 2008.
[18] A. van der ziel. Noise in solid-state devices and lasers. *Proceedings of IEEE*. 58(8):1178–1205, August 1970.
[19] Y. Isobe, K. Hara, D. Navarro, Y. Takeda, T. Ezaki and M. M. Mattausch. Shot noise modeling in Metal-Oxide-Semiconductor field Effect transistors under sub-threshold condition. *IEICE Trans on Electronics*, 90(4):885–894, January 2002.
[20] A. P. Huang, Z. C. Yang and Paul K. Chu. Hafnium-based high-k dielectrics. *Advanced in solid state circuit technologies*, April 2010.

Biographies

Wafa Makni was born in Sfax, Tunisia in 1980. She received the Electrical Engineering Diploma in 2004. In 2007, she obtained her Master degree in new technology of computer systems from the National School of Engineering of Sfax (ENIS). She joined the Electronic and Information Technology Laboratory of Sfax "LETI" and received her PhD degree in 2015 (ENIS). His current research interest is in carbon nanotube transistor modeling.

Montasar Najari was born in Tunis, Tunisia in 1981. He received bachelor diploma from the National Institute of Sciences and technology (Tunis - Tunisia), Master of science from the National School of Engineering (Sfax-Tunisia) and Ph.D form The University of Bordeaux 1 (Bordeaux-France) in Electronic engineering in 2004, 2006 and 2010 respectively. During his Ph.D. studies he had involved research in the topic of Nanoelectronics devices modeling and their circuit applications. He worked as a lecturer in The University of Bordeaux 1 (France) and Polytechnic National Institute of Grenoble (France) in 2010 and 2011 respectively before joining the University of Gabes (Tunisia) as an assistant professor from 2011 to 2015. Now Dr. Najari is consultant supervisor of the Innovation Unit (IKCE) in Jazan University (Saudi Arabia). His main topic of research is focusing on the modeling of post-silicon materials integration as carbon nanotubes and graphene in transistor devices and their digital / analog circuit application; i. e. logical inverter, ring oscillator, RAM cells, inion selective FETs, biosensors.

Hekmet Samet was born in Sfax (Tunisia) in 1956. She obtained the Engineering Diploma and a PhD in physical sciences from National school of Engineering of Sfax (ENIS) and the accreditation to supervise research in electronic, respectively in 1983, 1997 and 2010. Currently, she is an associate Professor at ENIS School of Engineering and a member in the "LETI" Laboratory.

R.-D. Berndt, M. C. Takenga, P. Preik, L. Berndt and S. Berndt

Validation of a SaaS-based Platform for Mobile Health Applications

Abstract: The demand for new healthcare services is growing rapidly. Involving ICT in health solutions has shown to raise satisfaction for both health care providers and patients. Several research works have been focusing on this issue, since it appears to be the suited solution for reaching an economically and socially viable solution to the increasing number of chronically ill patients. Although the development of ICT infrastructure gives the possibility to provide new services on a wider scale, some issues dealing with security, privacy, scalability and interoperability are still critical. Moreover changes in information flows, along with an explosion of digital content that needs to be stored and shared, are driving the need for a secure, flexible and scalable IT platform through which providers, payers and health sciences can support collaboration and information exchange. This paper introduces a secure and scalable platform which enables the implementation of mobile health applications and their provision based on SaaS (software as a service) technology. The platform has been validated through trials in the following fields: events monitoring in seniors' environments, diabetes management, teledermatology, fitness and Stress monitoring.

Keywords: mHealth, eHealth, Monitoring, Secure Platform, SaaS, Diabetes Management.

1 Introduction and state of the art

Modern information technology is increasingly used in healthcare with the goal to improve and enhance medical services and to reduce costs. In particular e-health systems like electronic health records are believed to decrease costs in healthcare (e. g. avoiding expensive double diagnoses, or repetitive drug administration) and to improve personal health management in general. E-health has been proved to be suitable candidate for the future of health care provision and it has become a very promising market for the industry. There is no wonder why big players such as Microsoft, Intel, Bosch and many others have been developing their eHealth platforms and related services in the last few years [1–3].

Mobile apps have enabled health institutions interact with their audiences in relevant ways than traditional channels. The next step in this evolution is securing this communication so that all types of content, especially protected health information

R.-D. Berndt, M. C. Takenga, P. Preik, L. Berndt and S. Berndt: IT-Company: Infokom GmbH Neubrandenburg, Germany. Emails: rberndt@infokom.de, ctakenga@infokom.de, ppreik@infokom.de, lberndt@infokom.de, sberndt@infokom.de.

De Gruyter Oldenbourg, ASSD – Advances in Systems, Signals and Devices, Volume 6, 2018, pp. 93–112.
https://doi.org/10.1515/9783110448375-007

can become part of the conversation. Security of this type cannot simply be part of the app, but rather it needs to be the cornerstone of an integrated communication platform that handles the data from beginning to end and all the way through the communication.

A number of projects have been focusing on the conception of platforms for the hosting of e-Health solutions in the last decades. In the year 1998, a telematics platform for patient oriented services was developed [4]. Three years later, a prototype for mobile telemedicine was conceived and presented in [5]; thereby the communication between the mobile phone and telemedical processor was enabled through the infrared (IrDA) interface which, however, does need a direct line of sight. In [6], a framework solution for information systems, which could be exploited for research projects in preventive medicine, is described.

The growing use of web-based user interfaces by applications continuously decreases the need for traditional client-server applications. This fact has motivated big players such as Oracle, IBM, SAP, Microsoft, Google, Amazon to react to the revolution of the SaaS technology [7]. They offer both SaaS services and Platform as a Service (PaaS) [8].

E-health systems store and process very sensitive data and should have a proper security and privacy framework and mechanisms since the disclosure of health data may have severe social consequences especially for patients. This paper introduces a modular and flexible platform suited for the hosting of mobile health solutions which are provided to end-users based on SaaS. This platform not only covers common functions such as authentication, authorization, secure communication and user management, but also supports flexible interfaces to hospital information systems (HIS) and easy and secure connection of mobile applications. Technical solutions for the protection of privacy-sensitive data which has not been appropriately addressed yet for end-user systems will be summarized. Moreover, the scalability and validation of the platform will be confirmed by conducted trials of multiple health applications implemented on it.

2 Implementation scenarios of mobile health applications

SaaS-based Platform can be applied for the implementation of mHealth applications that can help patients manage their treatments when attention from health workers is costly, unavailable, or difficult to obtain regularly. Better recordkeeping is another widespread outcome of mHealth technologies. Other mHealth applications are designed to capture real-time health information used to monitor the evolution of patients' states. A number of researches has paid attention on this topic [9–13].

Figure 1 illustrates the implementation scenarios of the four developed applications, namely FEALESS, MobilDiab, MobilSkin and eHealth-MV which resulted from previous conducted projects. They provide respectivelly solution for fall and fire detection in senior environments, diabetes management, teledermatology, fitness and stress monitoring.

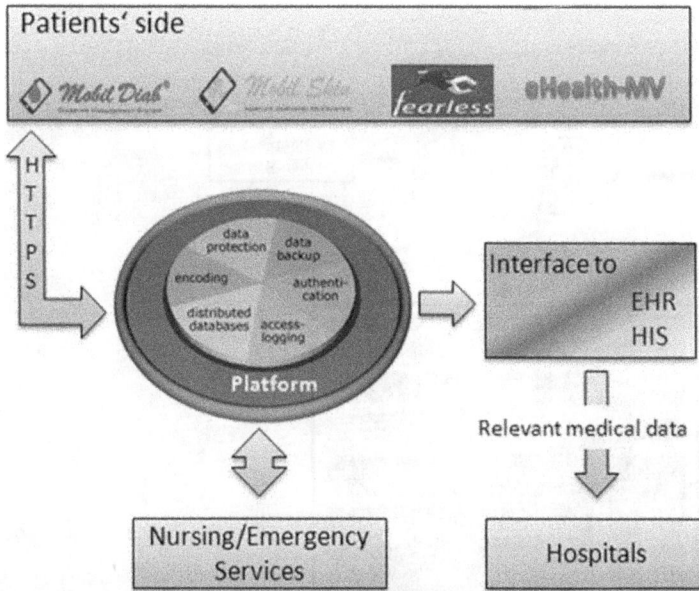

Fig. 1. Implementation scenario of health applications on the SaaS-Platform. EHR: Electronic Health Record. HIS: Hospital Information System.

2.1 FEARLESS implemented on the platform

Fear Elimination As Resolution for Loosing Elderly's Substantial Sorrows (FEARLESS) is a joint and ongoing project (AAL-JP) in cooperation with partners from four European countries (Austria, Italy, Spain and Germany) and consists in designing a system able to detect a wide range of risks (e. g. fall detection, smoke and fire) with a single sensor unit, enhancing mobility and enabling elderly to take active part in the self-serve society by reducing their fears [16].

FEARLES system is able to detect a number of events in the elderly home environments. These include falls, fire, smoke and unusual behavior in the daily activities. If a risk (i. e. fall, fire or smoke) is detected, an event is sent to the platform, as illustrated in Fig. 2. Depending on the confidence value provided within this message, different escalation levels are addressed. High confidence values requires immediate response to the event, lower confidences requires a verification or falsification by the call center agent. If an event is detected, a simplified illustration of the scene is transmitted to the call center.

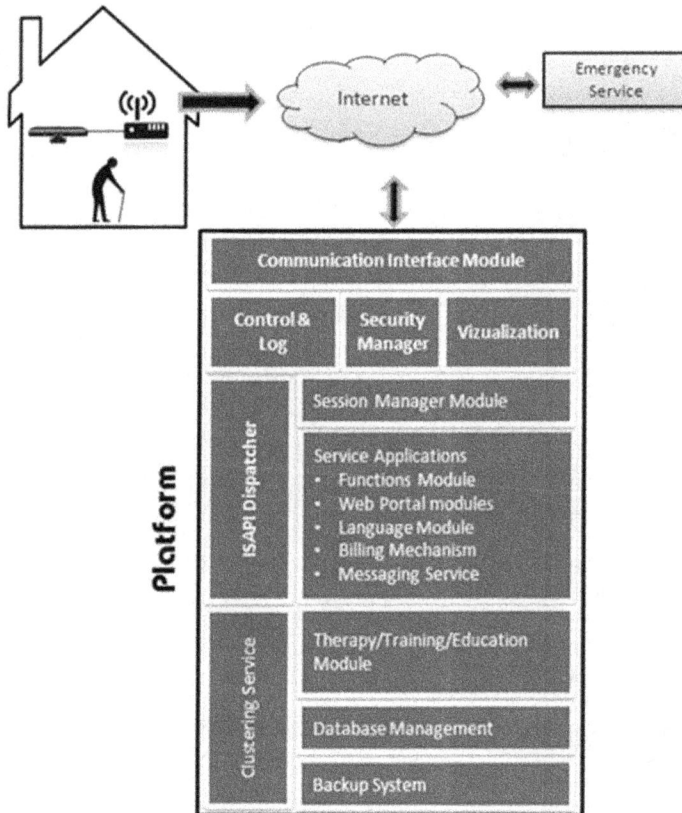

Fig. 2. FEARLESS System on the Platform.

The combined visual and acoustic detection of risks in a fusion approach served as basis for detecting falls events [16]. However, compared to standard IP cameras, the use of the kinect for Xbox/Asus Xtion Pro (Live) was proven to suit better for this project.

The analysis of images from the Kinect served also as input to the fire and smoke detection algorithm developed by the Fraunhofer-IPK. For fire detection, the following features were extracted: color analysis (reddish colors to white), fast stochastic changing contour or shape, fast changing lightness (high frequency luminance), color, geometry, and motion of fire region, moving or not moving target. For detection of the smoke, Contour based (grayish), changing or losing sharpness, periodic behavior are the features to be extracted for decision making [16].

The Platform fulfills in this project among others the following tasks: Enabling a secure communication and data exchange among partners, connection of web-based application for the user interaction to the system, connection and integration of health care services providers such as call center and emergency stations for intervention, secure communication interface to hospital information systems and to the electronic health record, access control and authentication of users.

The platform is can be illustrated as a set of modules, Fig. 2. The main module is the session manager which is responsible for the login process, session creation and encoding of data session-depended, and access control to the database. The ISAPI Dispatcher module is responsible for the communication with the Microsoft Web Server (Internet Information Service-IIS) in order to bring the services to the internet. Portal Modules are set of codes which execute the functionalities covered by web-portals. The platform enables an optimal load sharing through its clustering feature.

Trials for the validation of the Fearless system are being conducted in Germany, Austria, Italy and Spain. First results have shown positive impact of the system on the seniors' lifestyles in their home environments. They feel secure and this helps enhance their participation in social life and their mobility at home.

2.2 Mobile diabetes management system (MobilDiab) implemented on the platform

MobilDiab system is an innovative solution for the assistance and care of diabetes patients. It enables diabetes patients manage their self-controlled data around the clock using their mobile devices (Android, iPhone, iPads) and/or web-based applications. Medical care access patients' data through a protected web portal [17]. The system is embedded on the Platform which is responsible for the following tasks: Connection of web and mobile applications, user-hierarchical model (administration, hospital, doctor, patient), control of access, secure interface to Hospital Information Systems.

MobilDiab system is composed by a mobile application (Android/iOS), web-based portals for different users' category and the platform as illustrated in Fig. 3. This system provides a series of benefits for patients, health care personnel and for the health care system.

Benefits for the patients include among others: unimpeded patient mobility, data input via smart phones and/or via web, regular self-control of diabetes-related data enables the right care to be administrated at the right time, potential to improve care process and quality of service, improvement in patient's motivation through their involvement in the therapy process, reduced check-up frequency to doctors, use of mobile health technologies encourages diabetes patients to change their behavior/lifestyle and improve their health.

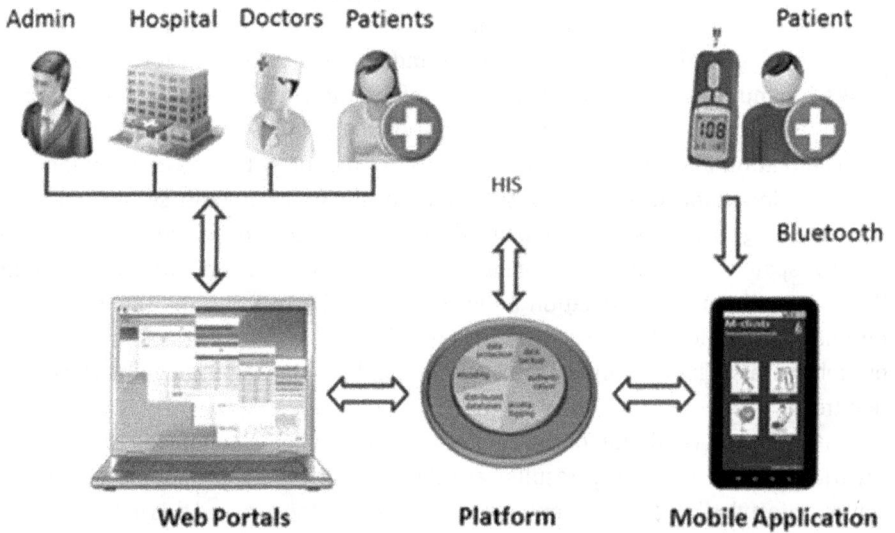

Fig. 3. MobilDiab System.

Benefits for the health care personnel involves among others the following: rich and regular data input which is helpful for individual therapy plan, minimization of errors caused by lack of information about the disease history, improvement of the care process quality, access to specialist opinions, access to patients' data worldwide independent from time and location, automatic alarm messaging in critical cases.

Benefits for the Health System include among others the following: delay and reduce diabetes complications, minimize hospitalization rates due to diabetes complications, reduce death rates from diabetes, speed up the transition of patients from hospitals to their own homes which leads to a reduction in costs, save health care expenses, potential to improve care process, enable the organization of health information through a centralized storage of patients' data, IT-enabled diabetes management with fully integrated provider-patient system. Features and design of the mobile application are presented in the screenshots of Figs. 4, 5. Patients can capture the blood glucose measures, nutrition information, pills to be taken or insulin to be injected, sport exercises done and photos related to the disease. Moreover, patients have the possibility to track blood pressure and weight. These parameters are of great importance for diabetes management.

As Output, the patient is given the choice to get his results being represented in a way he could get the statistical information (Fig. 5a), in a way he could track the trend and evolution of his disease (Fig. 5b, 5d) or in a way he could get important information for each day (Fig. 5c).

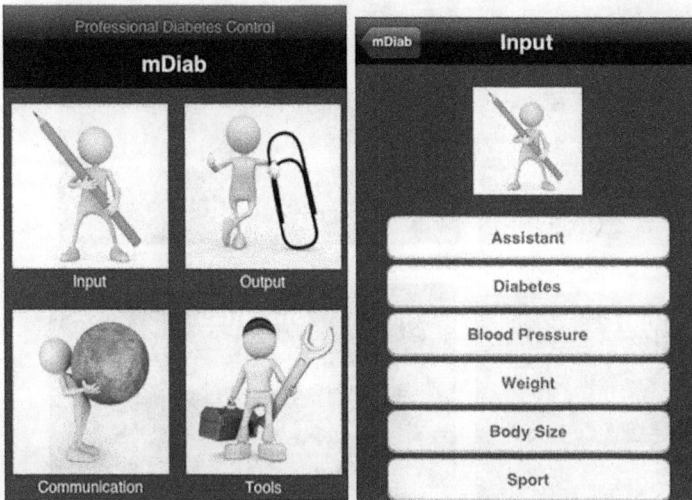

Fig. 4. mDiab- Input and Output Screens, iPhone screenhots (a) Main screen, (b) Input choices.

For the validation of the system, three studies with three different user categories (children, adults, and seniors) were conducted. The first study focused on the evaluation of the impact of using the mobile diabetes management system 'Mobil Diab' on the treatment and therapy of diabetic children and adolescents with type-1. The Flowchart of the clinical studies performed is illustrated on Fig. 6.

Sixty-eight Children and teens took part in the trial conducted during their medical-psychological rehabilitation period at the Diabetes Clinic in the North-Eastern region of Germany. An overall amelioration of the HbA1c values has

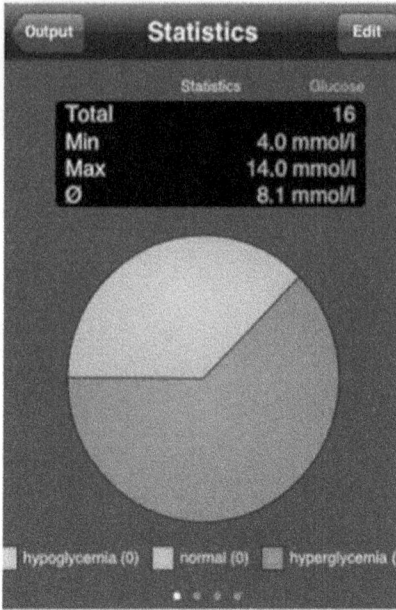

(a)

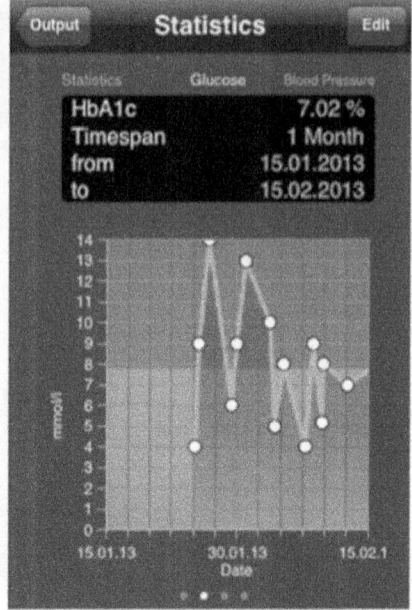

(b)

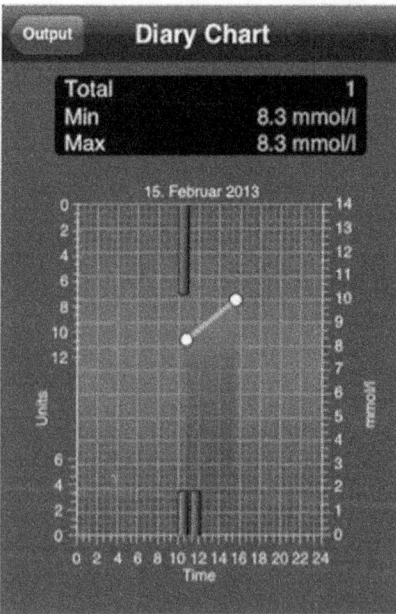

(c)

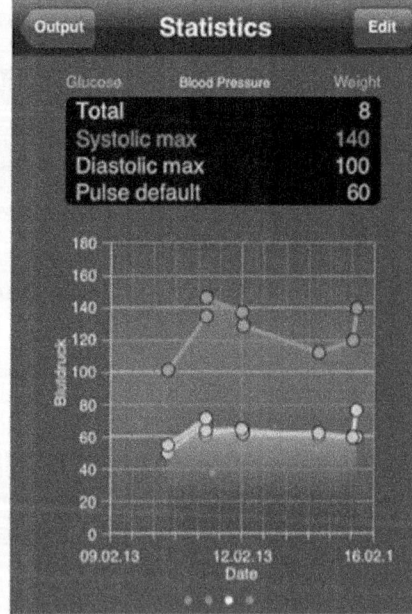

(d)

Fig. 5. mDiab- Output graphics, iPhone screenhots, (a) Statistics of blood glucose values, (b) Blood glucose trend, (c) Diary chart of blood glucose, insulin and carbohydrate, (d) Blood pressure graph.

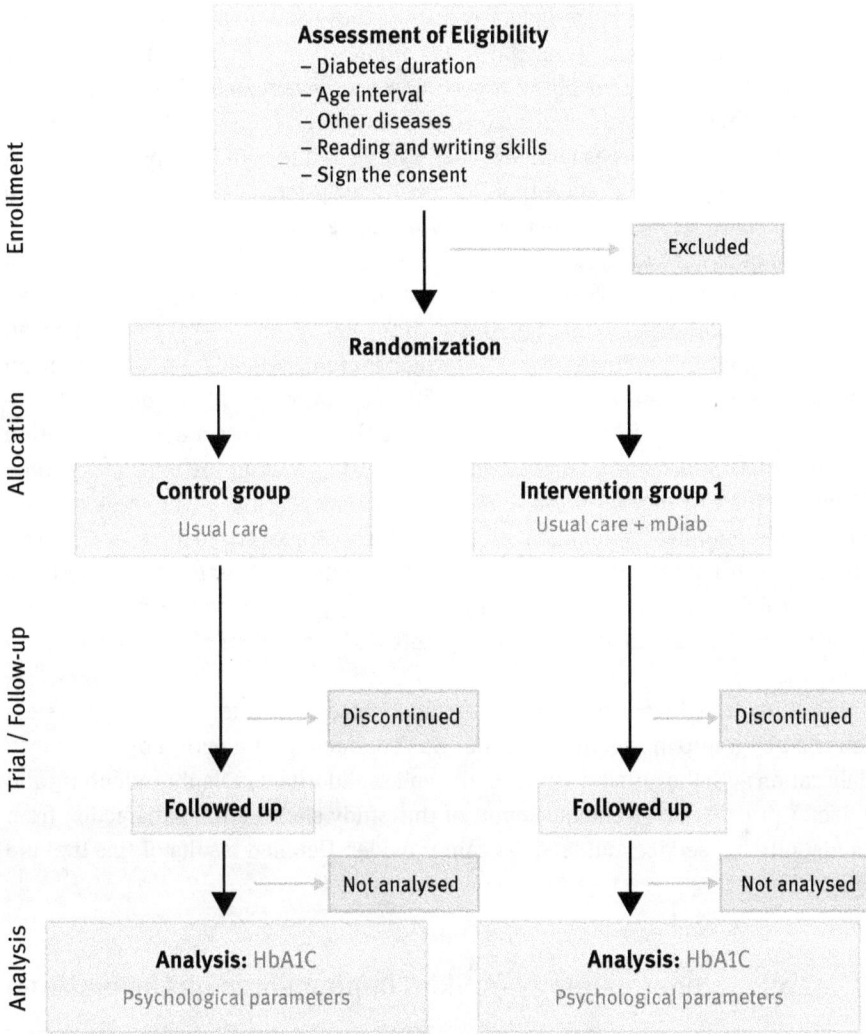

Fig. 6. Flowchart of the conducted clinical randomized studies.

been observed in both the control and the intervention groups; this demonstrated the positive impact of the medical-psychological rehabilitation on the average glycemic control. Moreover, results of the conducted trial demonstrated how the use of the mobile diabetes management system "Mobil Diab" obtained high patient and medical staff satisfaction scores [18].

The second trial aimed to test the effectiveness of the Diabetes Management System Mobil Diab in the context of African health care system. 40 adult patients coached by the health institutions of the health department of the Baptist Community in Central Africa (CBCA), were involved in the study conducted in the Democratic Republic of Congo. Results from the trial conducted in two eastern cities of the Democratic Republic of Congo have proved how the system Mobil Diab is suited not only for developed countries but also for communities traditionally underserved, those in remote or rural areas with few health services and staff. This population can access diabetes care using Mobil Diab, because the system overcomes distance and time barriers between health-care providers and patients.

The use of Mobil Diab showed to improve clinical outcomes of the patients from the intervention group involved in the conducted trial. This has been demonstrated from the amelioration of both the HbA1c (from 8.67 % to 6.89 %) and the mean amplitude of glycemic excursions (MAGE) which is characterized from both the mean blood glucose and its standard deviation. The decrease of the blood glucose fluctuations is demonstrated from results of the mean blood glucose standard deviation from the intervention group compared to the control group at the end of the study (33.0 mg/ml instead of 48.6 mg/ml). This proves how the use of the system could help patients stabilize better their glucose values. Moreover; positive evaluations of the system from patients and medical staff have been presented based on three metrics: usability and design, efficiency and therapy satisfaction, acceptance and appreciations. The obtained scores are 7 points and greater out of the 10 maximum points [19].

The third trial involved 10 seniors with diabetes, in order to test the system when used by a nursing service. The trial was conducted for a period of one year in collaboration with the nursing service of the Volkssolidaritaet in Neubrandenburg and the doctor practice involvement. Report of this study shows a full satisfaction from patients, nursing service and medical care provider. Detailed results of this trial are published in [20].

2.3 Mobile teledermatology (mSkin) implemented on the platform

The MobilSkin (mSkin) is an innovative mobile teledermatology system for supporting both the therapy and treatment of patients suffering from skin diseases, [14]. The system is based on the so-called store-and-forward teledermatology and consists of three main components: a mobile application (Android/iOS) to enable patients capture and send data related to the disease, a web-based portal for dermatologists and the whole system is hosted on the Platform.

In order to achieve a realistic and effective system for the targeted medical context, developers worked closely with dermatologists, nurses and patients throughout the whole project. Features and design of the application are shown in Figs. 7 and 8.

For the validation of the mSkin system, a trial was conducted in collaboration with the Clinic for skin diseases of the University of Greifswald in Germany. 13 Patients with skin diseases such as ulcus cruris, postoperative wound documentation, Eczema,

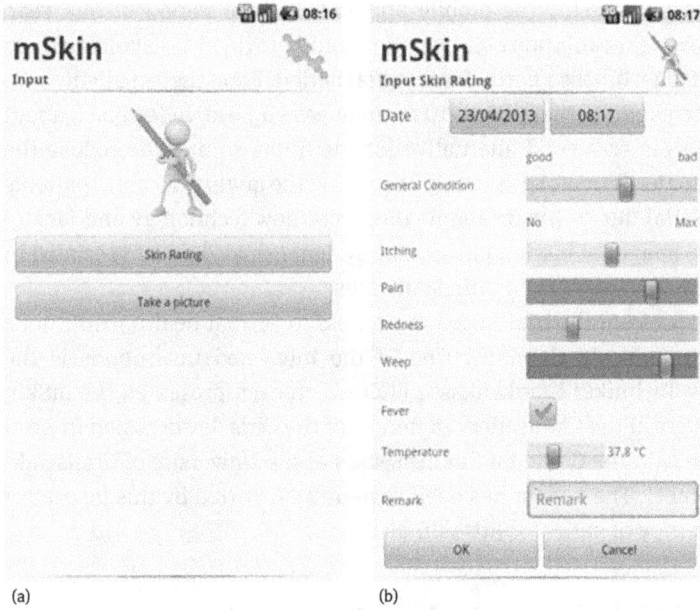

Fig. 7. mSkin, Input and Output screenshots, Android Samsung screenshot, (a)Input choice, (b) Skin rating.

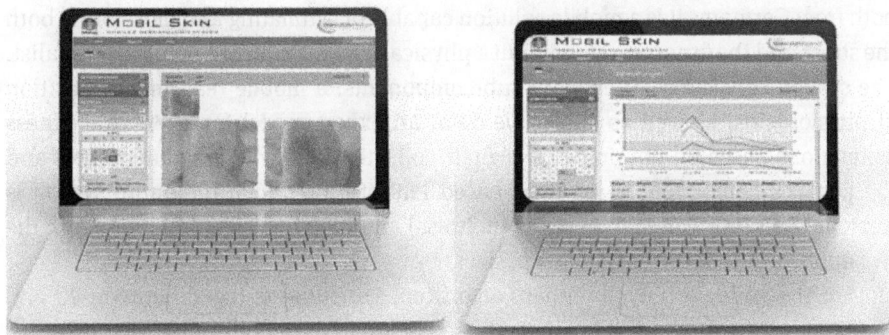

Fig. 8. mSkin, Doctors' web portal, Comparison and graphs pages.

Psoriasis and Pemphigus tested the mobile application for a period of one year. Dermatologists tracked the evolution of their patients and provided assistance to them via the online secured web-based portal for this trial period. From the 13 patients who were involved in the experimental phase, 10 of them gave a positive feedback about it. They saw the system as a good alternative for the future in order to reduce the frequency of visits to the dermatologist while improving the quality of care. 2 patients withdrew from the trial due to frustration dealing with new technology and lack of assistance in taking images at hidden places. The system can also assist specialist in the early detection of skin cancers and unusual cases.

Positive feedbacks from the trial raised awareness to several health institutions and insurance companies in Germany. One of the interested institutions is the insurance company Techniker Krankenkasse (TK) which even presented the mSkin technology and system at the Chancellery in Berlin for the Girls day occasion in April 2013 in cooperation with the Clinic for Skin Diseases of the University of Greifswald and the Infokom GmbH. The system has been currently supported by this insurance company for the therapy of patients with skin diseases.

3 eHealth-MV implemented on the platform

The eHealth-MV (eHealth-Mecklenburg Vorpommern) system was developed as the result of the cooperation between the University of Rostock (Institute for Preventive Medicine, Institute for Automation Technology) and the IT-Company Infokom GmbH, both from Germany. It is a mobile solution capable of estimating and monitoring both the stress and the fitness levels without a physical consultation of a medical specialist. The system consists of three main sub-components: a mobile real-time acquisition of physiological as well as subjective data, an expert model for stress and fitness estimations based on physiological signals collected from wireless vital sensors and the last sub-component is the SaaS-based Platform on which the entire system is implemented [15]. The mobile stress (mStress) and mobile fitness (mFitness) are the two applications developed in this project.

For the validation of the fitness algorithm, 110 subjects (aged between 20 and 56 years), whose fitness levels were assessed at different periods of the year, were involved in the experimental part. Subjects were guided through the testing procedures by a smart phone application. Heart rate and walking time as well as other input variables e. g. age and gender were automatically processed for an instantaneous fitness evaluation on the smart phone/server. Developed models could provide a high-correlation coefficients with directly measured oxygen consumption $r = 0.85$ and above with low standard error of estimate $4.3\,ml/min\;kg^{-1}$ and below. The obtained accuracy is sufficient for cardiorespiratory fitness classification in a general population. For the end user, Fitness levels were classified in seven levels, from an extremely bad to an excellent fitness level. Detailed information about this is described in [15].

The validation of the Stress estimation algorithm was done through a trial in which 50 different subjects took part. The aim was to monitor the stress level of the subjects over the time period of 24-hours. The stress prediction algorithm should be accepted if it results in a "good" correlation between the predicted values and subjective rating score of stress. The subjective ratings were obtained using the mobile handheld where the software version of modified NASA Task Load Index (TLX) was implemented. The mobile handheld was programmed to ask the subjects through dialogues for inputting the subjective rating score of stress felt by them during last 5 minutes. The subjective ratings of stress were collected for each subject at different times of the day whenever a certain level of change in the heart rate of the subject was detected. It was observed that the predicted stress values are positively correlated with the subjective rating score of stress with R = 0.7729, [15].

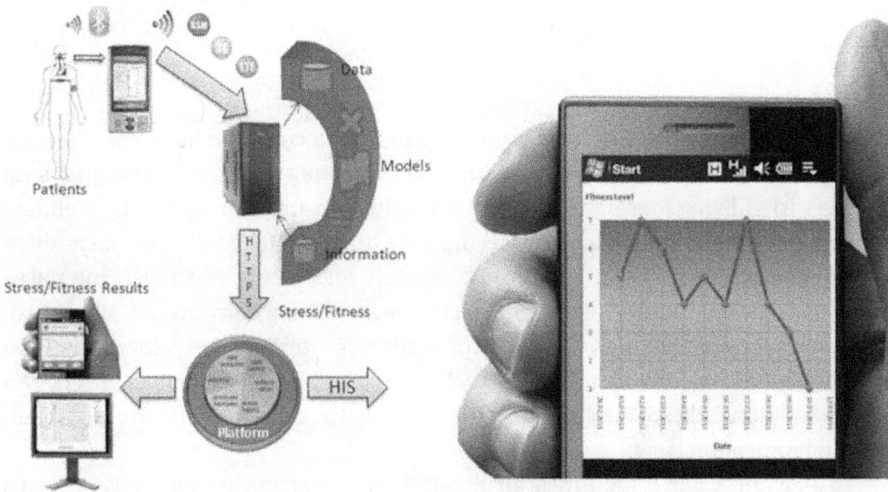

Fig. 9. (a): eHealth-MV System implemented on the platform, (b): mobile application, fitness state graph.

4 Architecture and basic functions of the platform

The patent-protected platform [21] illustrated in Fig. 10 has been developed to help bridge the gap between health and mobility. Diverse healthcare modules have been

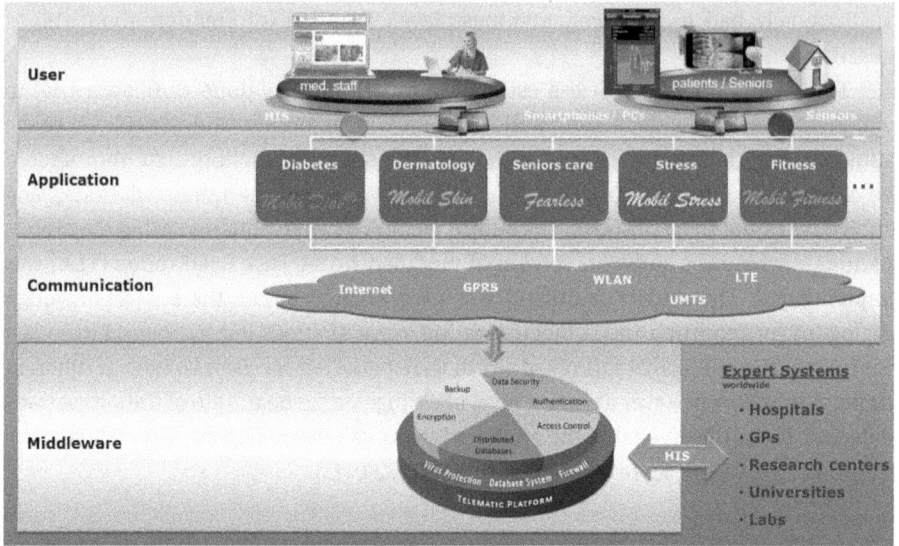

Fig. 10. Layered-structure of the Platform.

implemented and other are planned in order to meet a complete health care service package from diabetes, dermatology, stress, fitness, long-term health conditions up to the assisted-living for senior. The platform helps consumers track health, wellness and vital information using a highly secure infrastructure. It allows consumers share information with their health care professionals and family. Moreover, interfaces to hospital information systems and practice management software are supported. The four-layer architecture of the platform enables users securely share sensitive information. Using different devices such as smartphones, computers, users can access functionalities of applications supported by the core of the platform through the communication layer.

In e-health systems, security is an imperative requirement because those systems handle very sensitive data like medical and personal data.

Basic features such as user management, billing, security, encryption, authentication, authorization control, support of third party communication interfaces and secure data storage are supported in the core layer of the platform

The platform is composed of a set of modules. Each module is for a particular service or group of services responsible. These services can be consumed at the application layer by the end-users as SaaS-model or as modular business solution. The modularity of the services enables flexibility in the mobile application solution and efficient management. Basic features which are supported by the platform are detailed bellow:

4.1 User management

This function allows the SaaS-Povider to register/ update/ delete user information and to manage end-users by grouping them. Figure 11 illustrates a sample of service provision based on SaaS model for a medical case. For the illustrated scenario, the SaaS-Platform provides different services (e. g. Diabetes Management, teledermatology, fitness/stress management, etc.) to health institutions which could be hospitals, GP practices, rehabilitation centers etc... The health institution acquires an administration tool and can easily register doctors and patients who will benefit from the offered service. Depending on the offered services, different scenarios are possible.

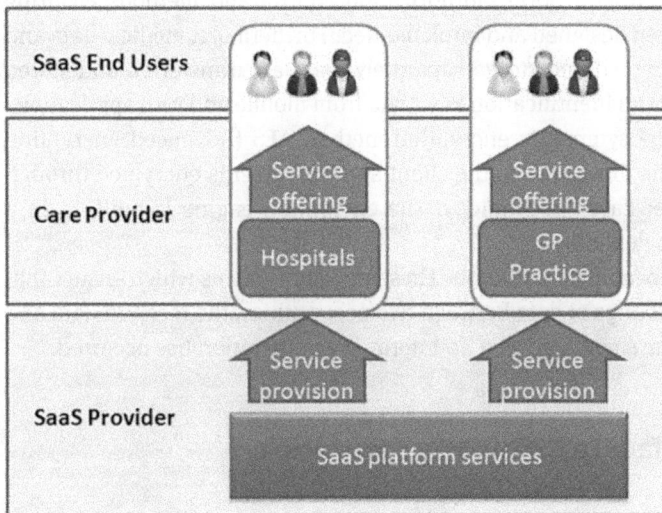

Fig. 11. Sample of a SaaS Model for a Medical Case.

4.2 Authentication

These are methods and mechanisms which allow an entity to prove its identity to a remote end.This function includes the mechanism that allows end users to authenticate themselves by entering a unique key by scanning the QR-Code containing their identification key.

4.3 Authorization control

The access control mechanisms and the ability of an entity to access shared resources. The platform includes an authorization mechanism that controls the access. This function sets the authorization to applications, services and some areas of the data according to the user category and privacy constraints.

4.4 Secure data communication and storage

During the conception phase of the platform, special consideration has been paid on the security and the privacy issues. Therefore, session models, high-level data encryption methods, authentication, authorization control and methods ensuring confidentiality have been designed and implemented. Furthermore, medical data and personal data are encrypted and stored separately. Data are transferred and stored anonymously with only an identification key. Data from mobile and web applications are encrypted using the symmetric encryption method AES (Advanced Encryption Standard). Moreover the link between the client and the server is encrypted through the SSL (Secure Socket Layer) technology. The encryption engine is built up in a modular way to enable flexibility.

Data integrity has been paid attention. These are mechanisms which ensure that when there is an interchange of data between two peer entities, the received data and the original ones are the same, and that no intermediate alteration has occurred.

4.5 Flexible interface to third-party application

The Platform is conceived in a way that it is able to support different types of Interfaces, including those based on XML and SOAP Standards. This alleviates the communication to third-party medical care providers and services that are using XML-based standards such as HL7/CDA (Health Level 7/ Clinic Document Architecture) and many others. Interfaces to most of the practice management software of the general practitioners, such as Doctor to Doctor (D2D) and KV-SafeNet are also supported by this Platform.

4.6 Connection of mobile and web applications

Mobile Applications of both Android and iOS operating systems (OS) can interact easily with the Platform as illustrated from the layered structure of the platform. A flexible web-portal hierarchy structured as Administration-Hospital-Doctor-Patient is supported by the platform. Such a structure enables an easy authorization control based on the user category. Encrypted data are transmitted anonymously through the SSL-based encrypted link.

5 Conclusion

The entire healthcare industry is undergoing a shift designed to enhance the level of care provided to patients. The sensitivity of patient information has created the need for high security solutions. This paper has presented a platform suited for secure implementation of mobile health applications. Basic functions and security features of the platform have been summarized. We have addressed the security, privacy and scalability issues. Proposed solutions for a secure communication, secure data storage, secure authorization and authentication method have been shortly presented. The developed platform has been successfully applied and tested for the implementation of mobile health applications offered to the market in the fields of ambient assisted living, diabetes management, teledermatology, stress and fitness monitoring. The conducted trials have successfully validated the system and have showed how end-users' fear on using e-health applications has been reduced, thus enabling high acceptance of SaaS-based services.

Bibliography

[1] *Microsoft-Health-Vault*. http://www.healthvault.com
[2] *Intel-Health-Guide* http://www.intel.com/corporate/healthcare/emea/eng/healthguide/index.htm
[3] *Bosch Health Care*. http://www.bosch-telemedizin.de/
[4] B. Zwantschko et al. Telematic Platform for Patient Oriented Services. *J. of Universal Computer Science*, 4:856–864, 1998.
[5] B. Woodward et al. Design of a Telemedicine System Using a Mobile Telephone. *IEEE Trans on Information Technology in Biomedicine*, 5:13–15, 2001.
[6] S. Holzmüller-Laue et al. A Highly Scalable Information System as Extendable Framework Solution for Medical R&D Projects. Editors: K. Adlassnig, B. Blobel, J. Mantas, I. Masic: Studies in Health Technology and Informatics: Medical Informatics in a United and Healthy Europe. *XXIInd Int. Congress of the European Federation for Medical Informatics*, (MIE'09), :101–105, 2009.
[7] P. Laird. *How Oracle, IBM, SAP, Microsoft, and Intuit are Responding to the SaaS*. http://en.wikipedia.org/wiki/Software_as_a_service
[8] *Platform as a Service*. http://en.wikipedia.org/wiki/Platform_-as_a_service
[9] C.R. Lyles, L.T., Harris, T. Le, J. Flowers, J. Tufano, D. Britt, J. Hoath, I.B. Hirsch, H.I. Goldberg and J.D. Ralston. Qualitative evaluation of a mobile phone and web-based collaborative care intervention for patients with type-2 diabetes. *Diabetes Technol. Ther*, 13:563–569, 2011.
[10] N. Tatara, E. Arsand, H. Nilsen and G. Hartvigsen. A review of mobile terminal-based applications for self-management of patients with diabetes. *Conf. on eHealth, Telemedicine, and Social Medicine*, eTELEMED'09, :166–175, Cancun, Mexico, February 1–7, 2009.

[11] I. Kouris, S. Mougiakakou, L. Scarnato, D. Iliopoulou, P. Diem, A. Vazeou and D. Koutsouris. Mobile phone technologies and advanced data analysis towards the enhancement of diabetes self-management. *Int. J. Electron. Healthc*, 5:386–402, 2010.

[12] A.Ramadas, K.F. Quek, C.K. Chan, B. Oldenburg. Web-based interventions for the management of type-2 diabetes mellitus: A systematic review of recent evidence. *Int. J. Med. Inform.*, 80:389–405, 2011.

[13] B.R. Raiff, J. Dallery. Internet-based contingency management to improve adherence with blood glucose testing recommendations for teens with type-1 diabetes. *J. Appl. Behav. Anal.*, 43:487–491, 2010.

[14] R.-D. Berndt, M. C. Takenga, S. Kuehn, P. Preik, D. Dubbermann, M. Juenger. Development of a mobile Teledermatology System. *J. of Telemedicine and e-Health*, 18(9), November, 2012.

[15] M. C. Takenga, R-D. Berndt, S. Kuehn, P. Preik, N. Stoll2, K. Thurow, M. Kumar, S. Behrendt, M. Weippert, A. Rieger, R. Stoll. Stress and Fitness Monitoring embedded on a modern Telematics Platform. *J. Telemedicine and e-Health*, 18(5):371–376, June, 2012.

[16] R-D. Berndt, M. C. Takenga, S. Kuehn, P. Preik, S. Berndt, M. Brandstoetter, R. Planinc, M. Kampel. An Assisted Living System for the Elderly - FEARLESS Concept. *Int. Conf., e-Health'12*, IADIS, :131–138, Lisbon, Portugal, July 17–23, 2012.

[17] R.-D. Berndt, M.C. Takenga, S. Kuehn, P. Preik, G. Sommer and S. Berndt. SaaS-platform for mobile health applications. 9th *Int. Multi-Conf. on Systems, Signals and Devices*, (SSD'12), Chemniz, Germany, March, 2012.

[18] R.-D. Berndt et al. Impact of Information Technology on the Therapy and Treatment of Type-1 Diabetes: Case Study with Diabetic Children and Adolescents in Germany. *J. Pers. Med.*, 4:200–217, 2014

[19] C. Takenga et al. An ICT-based Diabetes Management System tested for Health Care Delivery in the African Context. *Int. J. of Telemedicine and Applications*, ID 437307, 2014.

[20] R-D. Berndt et al. *Report of the study Mobil Diab: Diabetes management with the Volkssolidaritaet nursing service in Germany.* http://www.infokom.de,

[21] *Telemedicine System Especially for Chronic Diseases.* WIPO Patent WO/2008/043341, April 17, 2008.

Biographies

Rolf-Dietrich Berndt was graduated at the Chemnitz Technical University in Physics and Electronics in 1974 in Germany. He founded Infokom GmbH in 1991, which is an ICT-Company providing complex software solutions. Berndt is a qualified data security and privacy representative in Germany and managing director of the Infokom GmbH. His research interests focus on integrating new information and communication technologies in the field of healthcare and on developing mobile health applications aiming at the decrease of high costs stressing the health systems. He has served as coordinator of different national funded projects and has participated in a number of International funded projects.

Mbusa Claude Takenga received the BSc. and MSc. degrees in Radio engineering and Telecommunication from the Saint Petersburg State Technical University in 2001 and 2003 respectively. In 2007, he received his PhD. Degree in electrical engineering from the Leibniz University of Hannover with the Institute for Communication Technology. From 2008 to 2009, he worked as research fellow with the Rostock University in a project aiming to develop a telemedical system for estimating the stress and fitness states of a person. Since 2009, he has been with the R&D department of the Infokom GmbH in Neubrandenburg, Germany. His research interests are focused on conceiving and developing new telemedical and e-health solutions. He also serves as Visiting Professor for Mobile Applications in Austria with the Alpen-Adria Klagenfurt University and as Associate Professor for Telecommunications, Software Engineering and Programming at the State University of Ruwenzori in the Democratic Republic of Congo.

Petra Preik received the engineering degree (Dipl.-Ing. (FH)) in computer science and civil engineering from the University of Applied Sciences, Neubrandenbur-Germany in 2009. After her graduation, she joined the R&D team of the Infokom GmbH in Neubrandenburg, Germany. Her research interests have been devoted to topics such as conception and implementation of new telemedical and e-health solutions and programming mobile applications.

Luise Berndt received the B. A. degrees in Early Education from the University of Applied Sciences in Neubrandenburg, Germany in 2008.

Sebastian Berndt received the diploma of business administration from the University of Applied Sciences in Wismar, Germany in 2008. He is currently working as a project manager at the Infokom GmbH Neubrandenburg.

M. M. Ali, K. M. Al-Aubidy, A. M. Derbas and A. W. Al-Mutairi

GPRS-Based Remote Sensing and Teleoperation of a Mobile Robot

Abstract: The main objective of this research was to design and implement a remote sensing and monitoring system running on mobile robot with obstacle avoidance capability in unreachable area. A simple mobile robot prototype with onboard sensors has been designed and implemented to scan and monitor several variables in the surrounding environment. Teleoperation of such a mobile robot is a challenging task that requires an efficient interface and a reliable real-time robot control to avoid obstacles. The proposed system enables the user (base station) to send commands to the remote station (mobile robot), and receive scanned data and images from the environment through the internet and mobile DTMF signal. The proposed system hardware and software was implemented using PROTUS development software to obtain the suitable design parameters. Then, real experiments have been achieved to demonstrate the system performance including both the ultrasonic teleoperation of mobile robot navigation to avoid obstacles, and real-time sensing and monitoring in unreachable area.

Keywords: Mobile Robot, Robot navigation, Remote sensing and monitoring, Wireless sensor networks, Obstacles avoidance.

1 Introduction

There has been a tremendous increase of interest in mobile robots and their applications. One of these applications is using wireless mobile robots to detect several variables in the environment [1]. Using mobile robots equipped with sensors are becoming widely used, especially in environments where human involvement is limited, impossible, or dangerous. These robots can be used to perform some dangerous tasks that are difficult for human to do, especially in hazardous environments. Nowadays, internet-based teleoperation of mobile robots has opened new opportunities in long distance learning, resources sharing, and remote experimentation [2]. A global positioning system (GPS) has become an efficient tool in the civilian and military applications. GPS technology works under different weather conditions and across the world by any person if he has a GPS receiver [3, 4]. It provides users to track locations, objects, and even individuals in outdoor locations. However, GPS at indoor locations results poor performance due to its lack of ability to transmit and receive signals across concrete buildings [5]. In such a case, RF communication solution

M. M. Ali, K. M. Al-Aubidy, A. M. Derbas and A. W. Al-Mutairi: Faculty of Engineering, Philadelphia University, Jordan. Emails: moh_19592002@yahoo.co.uk, alaubidy@gmail.com.

De Gruyter Oldenbourg, ASSD – Advances in Systems, Signals and Devices, Volume 6, 2018, pp. 113–128.
https://doi.org/10.1515/9783110448375-008

is more suitable for indoor communication. In this project, the GPS technology has been used for mobile robot navigation and positioning in outdoor locations. Wireless Sensor Network (WSN) technology together with mobile robots can be used to detect several variables in the environment [6]. Each mobile robot can be considered as an individual node in the network which can monitor its local region and communicate through a wireless channel with other nodes to collaboratively produce a high-level representation of the environment's states. By using such a network, large areas can be monitored with low cost. There are many published papers combining wireless sensor networks with mobile robots. LaMarca et al. [7] used robots to increase the feasibility of WSNs, since sensor networks can acquire data but lack actuation, while robots have actuation but limited coverage in sensing. Rahimi et al. [8] describes a method of extending the lifetime of a wireless sensor network by exploiting mobile robots that move in search of energy, recharge, and deliver energy to immobile, energy-dependent nodes. Luo and Chen [9] demonstrated a remote supervisory control architecture which combines computer network and an autonomous mobile robot. A general purpose computer with internet access is required to command the mobile robot in a remote location through Internet. Hiroshi et al. [10] proposed a method to remotely locate a source using a mobile robot with gas sensors. The proposed method enables the estimation of the distance to the source together with its direction, the source can be remotely located. Wanga et al. [11] present an overview on recent development, future trends and advantages of wireless sensor technologies and standards for wireless communications as applied to wireless sensors. Mandow [12] proposed an autonomous mobile robot system equipped with appropriate sensors and operation devices to substitute hard and unhealthy human work inside greenhouses. Axaccia et al. [13] proposed a service robot for health monitoring and localized chemical, drugs and fertilizers dispensing to plants in greenhouses.

This paper presents the first stage of a project developed at Philadelphia University-Jordan. The main objective of the project is to use concepts of reverse engineering to design and implement a reliable and cheap communication channel running on mobile phone to guide a mobile robot with safe and efficient operation. The mobile robot can operate and perform scanning, monitoring and control tasks that are tedious and repetitive in a dangerous environment. The mobile robot is provided with microcontroller-based data acquisition unit and a real-time control and navigation algorithm to avoid obstacles.

2 Mobile robot design

The simplest type of mobile robots are wheeled robots that comprise one or more driven wheels and may have optional passive or caster wheels and possibly steered wheels [1]. The implemented mobile robot is a four-wheeled vehicle prototype with dimensions; 32 cm wide, 40 cm long and 25 cm height. Two DC motors are used for driving and steering, the first DC motor is used for driving both rear wheels via a differential box, while the second DC motor is used for combined steering of both front wheels, as shown in Fig. 1. The prototype provides support to the batteries, camera, and all elements related to the proposed design.

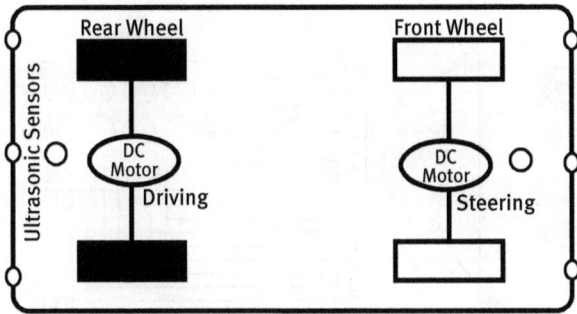

Fig. 1. Wheeled mobile robot.

(a). DC motor interfacing:
The first step when building a mobile robot is to select the appropriate motor and its control system. DC motors are the most commonly used in mobile robots, since they are clean, quiet, easy controlled, and can produce sufficient power. H-bridge circuit type (L293D) is needed to drive each DC motor forward and backward, as given in Fig. 2. Four output lines from the microcontroller are connected to the H-bridge to drive the DC motors.

(b). Ultrasonic sensors interfacing:
The mobile robot has eight built in ultrasonic sensors (type HC-SR04) to detect the route and avoid any obstacle in the working environment. Six sensors are distributed

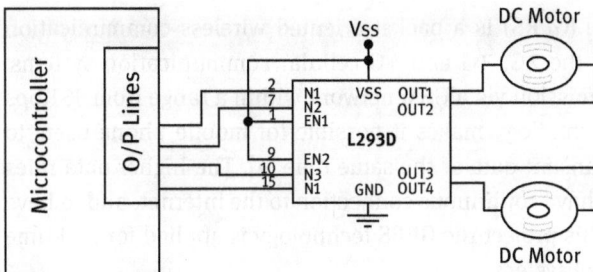

Fig. 2. Motor interfacing.

forward and backward to detect the route, while the other two sensors are fixed in the front middle and rear middle to detect any hole in its route, as shown in Fig. 3-a. Also, an internet camera is used to provide user the required broader area of sensing. The ultrasonic sensor includes an ultrasonic transmitter, receiver, and a simple control circuit. It provides 2 cm-400 cm non-contact measurement function, and covers good range accuracy (about 3 mm) with stable readings that are not affected by sunlight or black material [14].

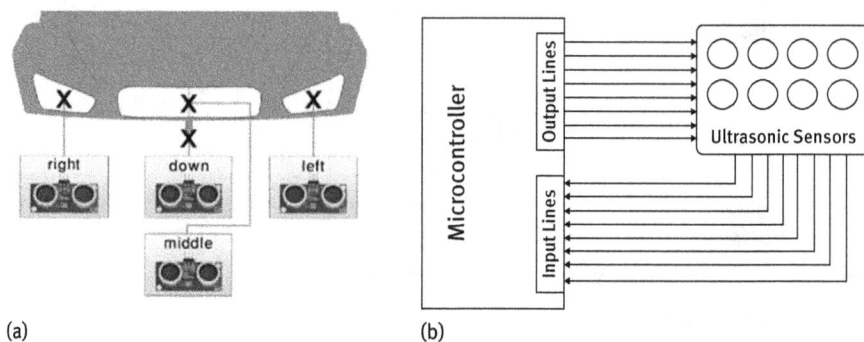

Fig. 3. Ultrasonic sensors. (a) Sensors distribution. (b). Interfacing with microcontroller.

To use this sensor, it is required to generate a short 10 μsec trigger pulse applied to the sensor input to start the ranging. The sensor will automatically send out an 8-cycle burst of ultrasound at 40 kHz and raise its echo. The Echo is a distance object that is pulse width and the range in proportion. In this case 16 input/output lines of the microcontroller are required to connect these sensors, as illustrated in Fig. 3-b. The internal timer/counter of the microcontroller is used to calculate the range through the time interval between sending trigger signal and receiving echo signal.

3 GPRS based communication modules

General packet radio service (GPRS) is a packet oriented wireless communication service available to users of the 2 G, 3 G and 4 G cellular communication systems. GPRS offers faster data transmission via a GSM network within a range from 56 Kbps up to 114 Kbps. The GPRS technology makes it possible for mobile phone users to make telephone calls and transmit data at the same time [4]. The higher data rates allow mobile phone users to have continuous connection to the internet, and to have web-based applications. In this project the GPRS technology is applied for real-time remote monitoring and control system.

The general layout of the proposed system is given in Fig. 4. The major components are the mobile robot, the microcontroller, the camera and two mobile phones. These components are mounted into two parts; the base station, and the remote station. The communication between base station and remote station takes place via mobile phone communication technology.

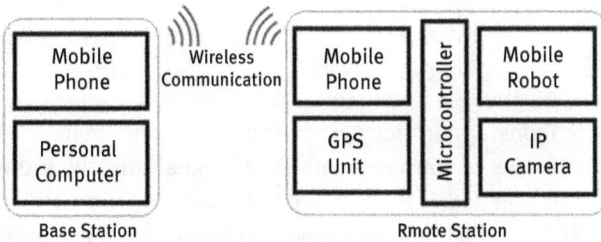

Fig. 4. General layout of the proposed system.

Base Station Design
The general layout of the base station architecture is shown in Fig. 5. It comprises a personal computer, a mobile phone, and an interface between personal computer and the mobile phone. The only way to access the mobile robot is to feed robot controller with commands from the base station mobile phone.

Computer Interface Unit
The input/output interface of the base station consists of a phone decoder, a phone generator and an interface card type Arduino UNO. It is a microcontroller-based board has 14 digital input/output pins, 6 analog inputs, a 16 MHz ceramic resonator, a USB connection, and a reset button. It contains everything needed to connect a mobile

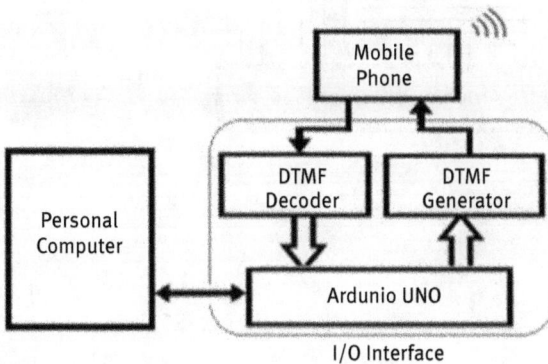

Fig. 5. Base station architecture.

phone to a computer with a USB cable [15]. The embedded microcontroller on the board is programmed to convert the mobile robot commands into an ON/OFF signals applied to the tone generator (TP5088) and then to the mobile phone of the base station. This tone is called dual-tone multiple-frequency (DTMF). Also, this interface converts the tone (send by the remote station through mobile) into a binary code using DTMF decoder (MT8870).

Tone Generator

The TP5080 DTMF generator provides low cost tone-dialing capability in microprocessor-controlled telephone applications. The 4-bit binary data generated from the base station microcontroller is converted into DTMF signal directly. If the tone enable input line is low, the 4-bit data is latched into the device and the selected tone pair from standard DTMF frequencies is generated. Figure 6 shows a typical interface circuit for direct generation of DTFM signal connected to the phone line of the base station.

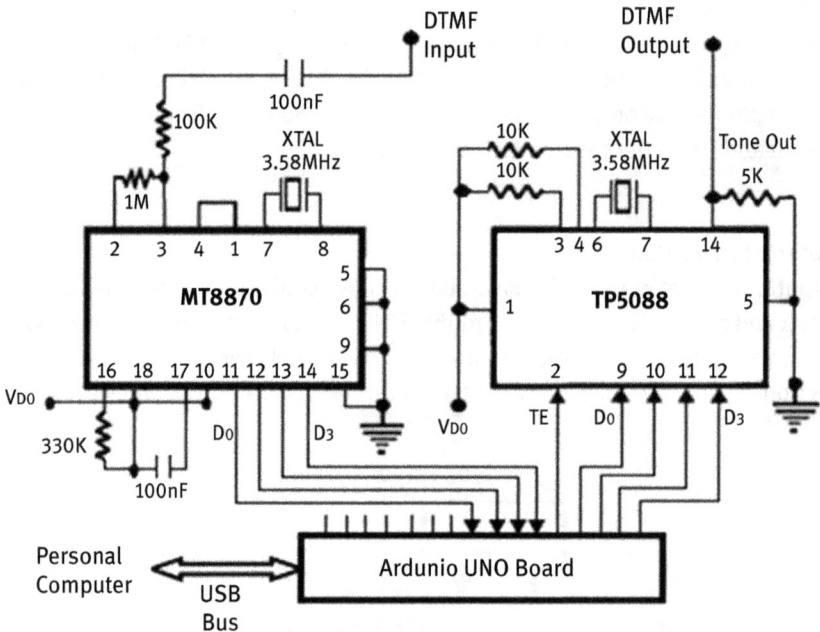

Fig. 6. Hardware design of the base station.

Tone Decoder

The DTMF decoder (type MT8870) is used for decoding the mobile DTMF tone signal received from the remote into 4-bit digital signal. The data acquisition algorithm of the base station converts the received 4-bit data into a digital signal representing the measured variable scanned by the remote station. The DTMF decoder is operated with a 3.579 MHz crystal. A capacitor of 100 nF is used to filter the noise and two resistors ($100\,K\Omega$ and $1\,M\Omega$) help to amplify the input signal using the internal amplifier, as illustrated in Fig. 6.

Remote Station Design

The general layout of the remote station is given in Fig. 7. It comprises a mobile robot, a microcontroller, set of environmental sensors, a mobile phone, a GPS receiver, a web camera, and an input/output interface between the microcontroller and the mobile phone.

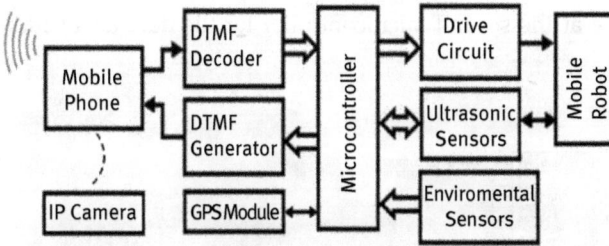

Fig. 7. Remote station architecture.

GPS Unit

The proposed remote monitoring and control system is based on satellite navigation technology, where a GPS is used to give the mobile robot the ability to determine its position. A GPS provides continuous positioning information, anywhere in the world under any weather conditions so long as the receiver has a direct line of sight to the sky [16]. A commercial GPS module type (SKM53) is used in this design due to its low power consumption which satisfies mobile applications. It has an embedded antenna which enables high performance navigation in the most stringent applications and solid fix even in harsh GPS visibility environments [17].

GPS module

The GPS module is mounted on the mobile robot, and interfaced to the microcontroller, as shown in Fig. 7. The data obtained from GPS receiver is processed by the base station to extract the mobile robot latitude and longitude values.

Microcontroller

The PIC16F877A microcontroller is used in this project. It is a low-power, high performance, 8-bit microcontroller with 14.3 KB flash memory, 256 bytes EEPROM data memory, 368 bytes of RAM used as internal data memory, 33 individually programmable I/O lines divided into five ports, 2 Comparators, 8 channels of 10-bit Analog-to-Digital converter, 2 capture/compare/PWM functions, 15 interrupt sources and a synchronous serial port.

Figure 8 illustrates the hardware design of the remote station, where all required units are connected to two microcontrollers. The total number of input/output lines required for the given design is 29 lines from the first microcontroller and 12 lines from the second microcontroller. Therefore, there are 6 analog input lines and 15 digital input/output lines available at the second microcontroller for the data acquisition unit.

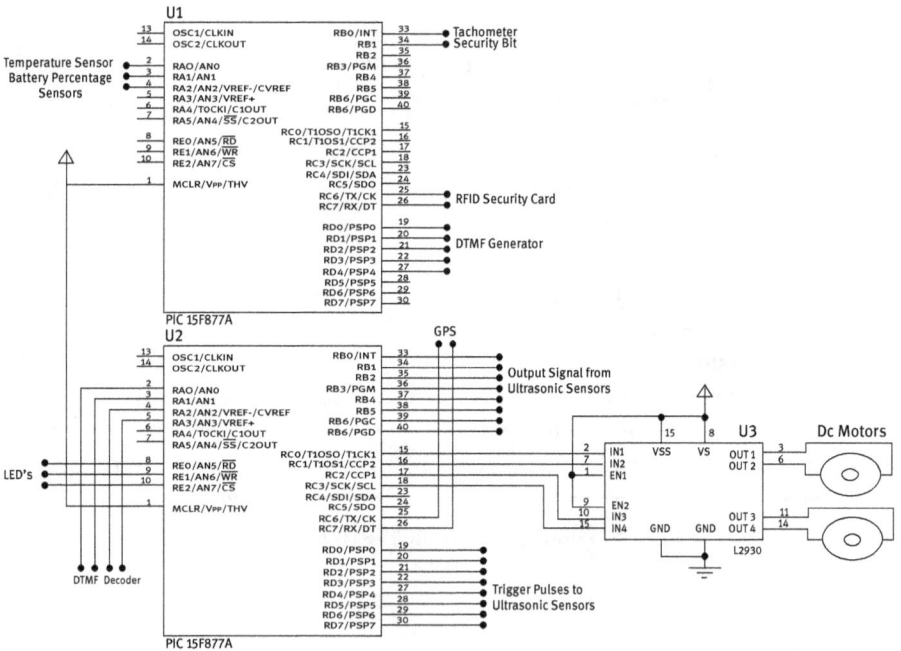

Fig. 8. Hardware design of the remote station.

Tone Decoder

Another DTMF decoder is required in the remote station for decoding the received mobile signal. It gets DTMF tone from the mobile headset's wire and decodes it into 4-bit digital signal to control the mobile robot. The DTMF decoder is operated with a 3.579 MHz crystal.

Tone Generator

the remote station has a tone generator similar to that used in the base station. The 4-bit binary data generated from the remote station microcontroller is converted directly to DTMF signal.

Data Acquisition Unit

The mobile robot has two groups of sensors; the first group is composed by eight Ultrasonic sensors which are mounted on the robot as illustrated in Fig. 1. The second group is composed by set of sensors to scan and monitor selected variables in the surrounding environment. There are 21 input/output lines available for this unit to interface the required sensors to the second microcontroller.

IP Camera

An internet camera is used as a remote sensing sensor to monitor the working area of the remote station. Requirements such as energy consumption, size and weight are very important in such application. The integrated wireless camera type "Tenvis JPT3815W" is used in this design. It provides 12 meters night vision functions, high quality video and two-way audio monitoring. It supports remote viewing and recording from anywhere anytime via web browser [5]. The IP camera uses WiFi technology, so it can be connected to a router and accessed across the internet.

Challenges of Implementation

1. As for the design of the DTMF circuit, there are ICs in market that do this job but with low quality the main reason is related to crystal inaccuracy besides power supply harmonics thus the detection especially when voice received is not very clear.

2. Motor driving was one of biggest troubles we faced, since of heavy robot weight so we had to find a way to drive motor without using mechanical parts such as relays. Once correct motor driver was found we had troubles with harmonics because of high starting current of motor, it forces controller board to reset, at same time we did not want to use a costly driver to keep design as cheap as possible.

3. Design and implementation of wireless charging unit in robot took a lot of trials, each time we lost something to work and mostly the MOSFET transistors cause of overheating, at end we synchronized circuit without using expensive parts like function generator.
4. Isolating controller from electromagnetic fields took some time, it is not because of the design issue but we were testing the robot inside strong magnetic fields to see if it can resist.
5. Programming was a challenge also; we wanted to design a stable control algorithm for robot with fast operation to serve real time needs.

4 Mobile robot teleoperation

Controlling mobile robots through teleoperation is a real challenging task that demands a flexible and efficient user interface. In this research, the mobile robot is equipped with numerous sensors including ultrasonic sensors, environmental sensors, positioning sensor and an IP camera. These sensors provide a high volume of data to the user, and the received information from the robot are used by the base station. In fact, teleoperations are necessary in such application where mobile robot is not able to deal with a certain task autonomously.

The mobile robot is controlled by a mobile phone (base station) that makes a call to the mobile phone attached to the robot. During a call, if any button is pressed, a tone corresponding to the button pressed is heard at the remote station. This tone is converted into binary code using DTMF decoder. The microcontroller is preprogrammed to take a decision for any given input and generates its decision to drive circuits of the driving and steering motors, as illustrated in Fig. 9.

Manual mode, the mobile robot reacts directly to the user commands. The user can guide the mobile robot to move forward, backward, turn left or turn right. If any obstacle is found, the robot will be stopped. Then, through the real-time images from the IP camera, the user will send the required commands to avoid obstacles. While, in the automatic mode, the ultrasonic sensors are used to assess the working environment in which the robot is moving. Upon detection of an obstacle or a hole by any of these sensors an action is taken by the control algorithm to avoid collision with any obstacle in the environment. Here the statuses of the sensors are read by the microcontroller to monitor the environment surrounding the robot.

5 Mobile robot navigation

Localization is the most important task for robot navigation and control. It is very important to know the robot position and its orientation at all times to make sure it reaches the final destination and covers the scanned area. This localization issue can be solved by using a global positioning system for outdoor scanning. In the case of indoor scanning, a GPS with infrared, sonar, laser or radio beacons can be used [1]. The driving system can update their position and orientation from time to time.

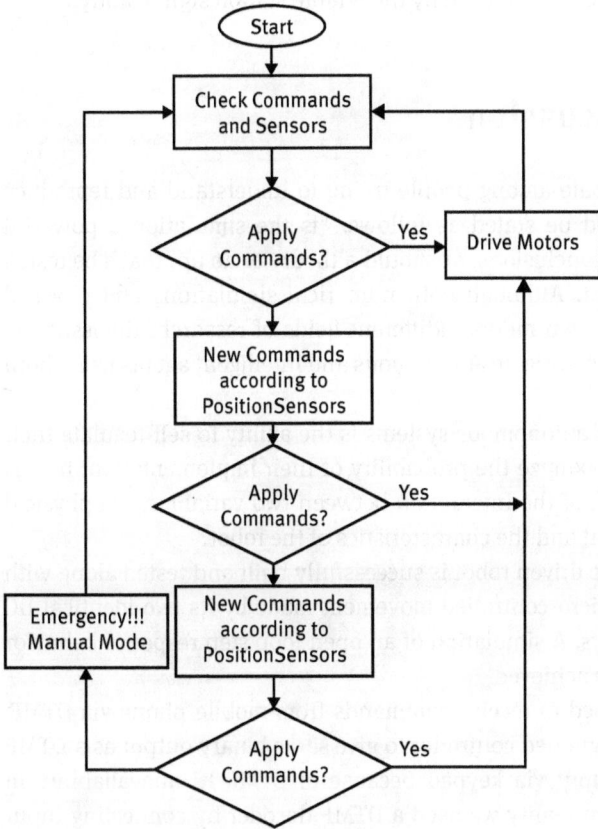

Fig. 9. Mobile robot navigation algorithm.

For remote control of Mobile robot, a rule-based control algorithm is used for obstacle avoidance. The IR sensors provided on front and back of the mobile robot are used for this purpose. The ultrasonic sensors can detect the obstacles at a distance of 400 cm from the robot. So these sensors provide the robot controller with the necessary information before collision takes place.

For the teleoperation, the mobile robot takes images of the remote environment and sends the visual information back to the base station over the Internet. In

sometimes it is not possible to watch certain regions due to the limitations of communication bandwidth and narrow view-angles of camera. To avoid this problem, new commands are generated from the base station to adjust the position of the IP camera and to move the mobile robot to new position according to the ultrasonic sensor data.

It is demonstrated by experiments that such actions improves the performance of the sensing and monitoring tasks achieved by the remote station significantly.

6 Results and discussion

There is currently a hot debate among people trying to understand and reproduce intelligent agents that could be stated as follows: "is the simulation a powerful enough tool to draw sound conclusions, or should a theory or an approach be tested on a real agent" , i.e. robot. Although both numerical simulations and physical implementations have their own merits indifferent fields of research, the issue becomes important when one investigate autonomous and intelligent agents using both ways [19].

A basic characteristic of autonomous systems is the ability to self-regulate their own behavior in order to maximize the probability of their implementation. In this sense adaptation is function of the interaction between two variables, the physical properties of the environment and the characteristics of the robot.

In this research a mobile driven robot is successfully built and tested along with a simulation study for its micro-controlled movement made by its two identical DC (Driving and Steering) motors. A simulation of an open loop step response behavior fit to the D.C motors model is achieved.

The robot is also designed to receive commands from mobile phone via DTMF, in simulation we used a programed controller to give same binary output as a DTMF decoder when it receives input via keypad because of DTMF IC unavailability in simulation software, while in reality we used a DTMF decoder by connecting input to mobile phone, and output to controller, results where good with reasonable delay between sending command and actuating it when using a 3G mobile network, vice versa for process is used with DTMF generator.

Our GPS was programmed in simulation to give a random position for check purposes, while in reality we used skylab GPS that connects serially to controller, it gives very accurate position when there is no coverage above it but the sky.

As for battery charge, we used voltage divider rule, which consists of two resistors to divide voltage, and input that voltage to controller. The controller knows maximum and minimum battery voltage while full and empty, it can calculate battery charge percentage and use that value to know the remaining operating time.

Results shown below reflect exactly what the mobile driven robot is designed for. Figure 10 shows the forward response of the driving motor subjected to an actuation ON signal, which clarifies its fast response to get the desired steady movement. The same conclusion is made when it received an OFF actuation order as in Fig. 11.

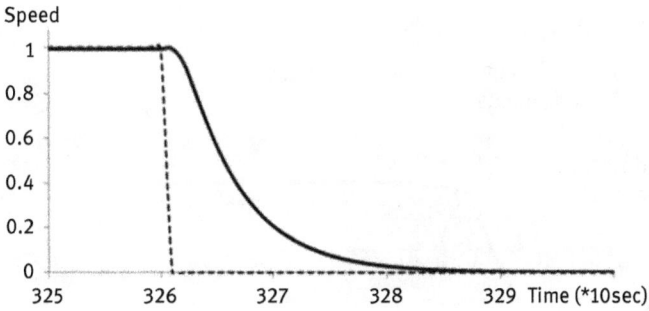

Fig. 10. Driving motor response for forward command.

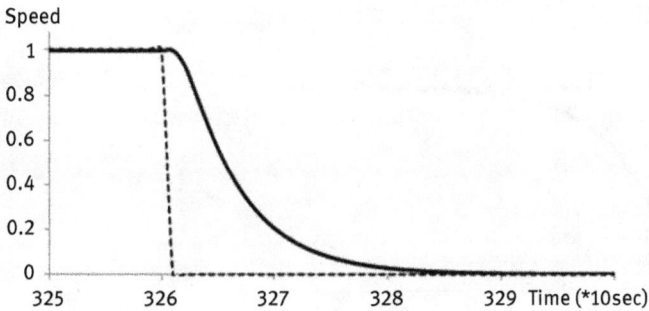

Fig. 11. Driving motor response for stop command.

As mentioned before that the ability of intelligent behavior of avoiding obstacles or getting rid of entering a hole is tested both in hardware implementation and simulation. Figure 12-a shows the ON-OFF-ON signals subjected to the driving motor at a certain period. It clarifies that when the forward motor is in its steady state motion and suddenly sensing a Hole the controller sends an OFF signal to this motor and after a while actuates it to ON state again as illustrated in Fig. 12-b. In the mean while an actuating signal triggers the stand still state of the Steering motor synchronized with the Driving motor late OFF response as shown in Fig. 12-c. The combination of both ON motors responses will avoid getting into a Hole or hit obstacles with a very enough safety period of time.

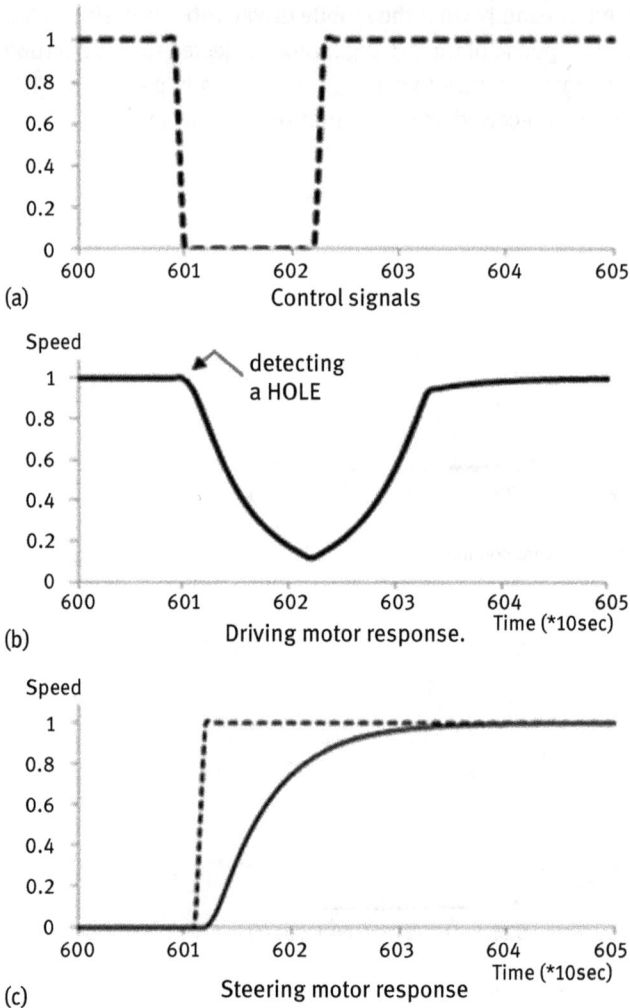

(a) Control signals

(b) Driving motor response.

detecting a HOLE

Time (*10sec)

(c) Steering motor response

Time (*10sec)

Fig. 12. Motors responses when detecting a hole.

7 Conclusion

This paper describes the first stage of a project developed at Philadelphia University-Jordan. A reliable and cheap communication channel running on mobile phone has been designed and implemented to guide a mobile robot equipped with data acquisition unit for remote sensing and monitoring system. Such a system can be used for real-time scanning and monitoring of several variables located in environments where human involvement is limited or dangerous.

The designed robot satisfies all basic planned goals that this research seeks for. Hardware prototype is firstly designed and verified using PROTUS workspace along with simulation results for the designed robot movements. These results make us confident in thinking that the underlying design will be encouraging to build a more complex mobile-driven requirements robot later on.

Bibliography

[1]   T. Braunl. *Embedded Robotics, mobile robot design and applications with embedded systems.* 2nd Edition, Springer, Germany 2006.
[2]   X. Xue, S.X. Yang and M.Q.-H Meng. Remote Sensing and Teleoperation of A Mobile Robot Via the Internet. *IEEE Int. Conf. on Information Acquisition*, July 2005.
[3]   *Engineers garage, How to interface GPS with PIC18F4550 Microcontroller.* http://www.engineersgarage.com/articles/global-positioning-system-gps.
[4]   *Cellular-news, What is GPRS?* http://www.cellular-news.com/gprs/what_is_gprs.shtml.
[5]   V. Nakka and A. Kabirdas. Design and realization of augmented reality based navigation assistance system. *Int. J. of computer science and information technologies*, 2(6):2842–2846, 2011.
[6]   G. Amato, M. Broxvall, S. Chessa, M. Dragone, C. Gennaro and C. Vairo. When wireless sensor networks meet robots. 7th *Int. Conf. on Systems & Networks Communication*, (ICSNC), :35–40, 2012.
[7]   A. LaMarca, W. Brunette, D. Koizumi, M. Lease, S.B. Sigurdsson, K. Sikorski, D. Fox and G. Borriello. *Plantcare: an investigation in practical ubiquitous systems.* 4th *Int. Conf. on Ubiquitous Computing*, :316–332, UK 2002.
[8]   M. Rahimi, H. Shah, G. Sukhatme, J. Heideman and D. Estrin. Studying the feasability of energy harvesting in a mobile sensor network. *IEEE Int. Conf. on Robotics & Automation*, (ICRA03), :19–24, September 2003.
[9]   R.C. Luo and T.M. Chen. Remote supervisory control of a sensor based mobile robot via Internet. *Int. Conf. on Intelligent Robots and Systems*, (IROS97), 1997.
[10] H. Ishida, T. Nakamoto and T. Moriizumi. Remote sensing of gas/odor source location and concentration distribution using mobile systems. Sensors and Actuators. *B: Chemical*, 49(1–2):52–57, June, 1998.
[11] N. Wanga, N. Zhangb and M. Wangc. Wireless sensors in agriculture and food industry-Recent development and future perspective. *Computers and Electronics in Agriculture*, 50(1):1–14, January, 2006.
[12] A. Mandow. The autonomous mobile robot AURORA for greenhouse operation. *IEEE Robotics & Automation Magazine*, 3(4):18–28, December 1996.
[13] G.M. Acaccia, R.C. Michelini, R.M. Molfino and R.P. Razzoli. Mobile robots in greenhouse cultivation: inspection and treatment of plants. 1st *Int. Workshop on Advances in Service Robotics*, Italy, 13–15 March, 2003.
[14] *ELEC Freaks, Ultrasonic Ranging Module HC-SR04.* http://elecfreaks.com/store.
[15] *Arduino, I/O Interface Board (Arduino UNO). http://arduino.cc/en/Main/arduinoBoardUno.*
[16] S. V. Ragavan and V. Ganapathy. A unified framework for robust conflict free robot navigation. *World academy of science, Engineering & Technology*, 25:88–94, 2007.
[17] *SKYLAB, SkyNav SKM53 Series, Ultra High Sensitivity and Low Power, The Smart Antenna GPS Module. http://ar.scribd.com/doc.*
[18] *Tenvis, IP Camera data sheet. http://www.tenvis.com/jpt3815w-hot-pantilt-ip-camera.*

[19] D. Floreano and F. Mondada. *Automatic Creation of an Autonomous Agent: Genetic Evolution of a Neural-Network Driven Robot.* 3rd *Int. Conf on Simulation of adaptive behavior*, SAB94, USA, :421–430, 1994.

Biographies

Mohammed M. Ali received his BSc, MSc, and Phd degrees in computer and control engineering from The University of Technology, Iraq in 1981, 1993, and 1998, respectively. he is currently an assistant professor in the Department of Computer Engineering at Philadelphia university, Jordan. His research interests include fuzzy logic, neural networks, image processing, and computer interfacing. he has 12 published papers related to neuro-fuzzy and real-time computer control applications.

Kasim M. Al-Aubidy received his BSc and MSc degree in control and computer engineering from the University of Technology, Irag in 1979 and 1982, respectively, and PhD degree in real-time computing from the University of Liverpool, England in 1989. He is currently a professor and dean of Engineering Faculty at Philadelphia University, Jordan. His research interests include fuzzy logic, neural networks, genetic algorithm and their real-time applications. He was the winner of Philadelphia Award for the best researcher in 2000. He is also the chief editor of two international journals, and a member of editorial board of several scientific journals. He has co-authored 4 books and published 76 papers on topics related to computer applications.

Ahmad M. Derbas was born in Amman, Jordan, in 1991. He received the Bachelor's and M.Sc. (with Honors) degrees in Computer engineering and Mechatronics engineering both from Philadelphia University, Jordan, in 2014 and 2017 respectively. He is currently a teacher assistant in the department of computer engineering at Philadelphia University, Jordan. His research interest includes embedded real time systems, and robotics. He published seven papers related to his field of interest.

Abdullah W. Al-Mutairi was born in Amman, Jordan, in 1993. He received the Bachelor's degree in Mechatronics engineering from Philadelphia University, Jordan, in 2016. He currently works as a Mechatronics engineer in Jakarta, Indonesia. His research interest includes embedded systems and Robotics. He has 4 published papers related to Robotics and automation.

M. S. Ben Ameur, A. Sakly and A. Mtibaa

A Hardware Implementation of Genetic Algorithms using FPGA Technology

Abstract: The main objective of this paper is an implementation written in Vhsic Hardware Description Language (VHDL) and intended for a hardware implementation for real time applications like Genetic Algorithms (GAs) which are one of the most advanced optimization techniques based on the concepts of natural selection and genetics. This paper presents a new FPGA-based structure of the hardware design using hardware implementation of GA which is robust parallel calculation method. GAs encode a potential solution to a specific problem on a simple chromosome like data structure and applies recombination operators to these structures like crossover and mutation. A data flow and a block diagram design are shown and described in the paper. Results demonstrate the requirements (logical blocks) needed for implementation for some various examples running at 50 MHz clock frequency. The inherent parallelism of these hardware implementation makes the running time negligible regardless the complexity of the processing.

Keywords: Genetic Algorithm, FPGA, Finite state machine, parallel programming, hardware implementation.

1 Introduction

Genetic algorithms have an important revolution, particularly in the using of Field-Programmable Gate Arrays (FPGAs) which are used as new digital solutions for the implementation of Genetic Algorithms (GAs). The first implementations of GAs were performed using microprocessors, microcontrollers and DSPs. These digital solutions have solved some design problems related to the use of linear functions, but did not give an optimal level optimization at execution time.

In this paper, we present the implementation of a genetic algorithm using FPGA. The body of this paper is presented as follow: In the first section, we present the mechanisms of a genetic algorithm. In particular, we present operators of selection, crossover and mutation and we are interested in the various methods of generating pseudo-random numbers. The objective of the next section is to describe, first, the overall functional architecture of the code perform the genetic algorithm. Second, we present the methodology and tools used in the simulation of some functions. In section 3, the GA is tested on several multimodal functions and it presents the

M. S. Ben Ameur, A. Sakly and A. Mtibaa: Laboratory of Electronic and Microelectronic, National Engineering School of Monastir, University of Monastir, Tunisia. Emails: msba2014@gmail.com, Sakly_anis@yahoo.fr, Abdellatif.mtibaa@enim.rnu.tn.

De Gruyter Oldenbourg, ASSD – Advances in Systems, Signals and Devices, Volume 6, 2018, pp. 129–144.
https://doi.org/10.1515/9783110448375-009

experimental results of some benchmark functions applied into the GA. Finally, some concluding remarks are summarized and some examples are drawn in section 4.

2 Genetic algorithms overview

In this section we present the mechanisms of GAs which is adapted for a quick exploration of a large search space and able to provide a several solutions. That's why using FPGA allows us to improve the performance of a given algorithm using the parallelism in a reduced portion of time. So the purpose of a genetic algorithm is to find the minimum of a defined function [1]. To implement a GA, we must begin with random population [2]. The method that generates a random number must be able to create a population of chromosomes not uniform or ordered to be used for future generations. After that, we evaluate these structures to allocate opportunities for chromosomes that represent a better solution to the target problem which gives more chances to reproduce than those chromosomes which are poorer solutions [3]. The selection of the initial population is very important because it can make faster convergence to global optimum. After selection operator we use crossover and mutation operator [4, 5] to modify the population over generations. The crossover operator reconstructs new chromosomes from the previous existing population and mutation operator ensures the exploration of the whole state space. Before starting our algorithm we must set some parameters as follows:
- The number of chromosome in the population
- The initial seed for the pseudo random number generator
- The type of selection operator
- The type of crossover operator
- The mutation probability
- The fitness of the chromosome in the population
- The number of iterations in a generated population

3 Hardware implementation of GA

To optimize a problem we have to explore all the search space in order to maximize (or minimize) a given function. The use of a genetic algorithm is suitable for a quick exploration [6] of an area. In this paper we present two types of encoding, binary encoding and real encoding. Each one has advantages and disadvantages compared to the others. Similarly, each encoding is more suitable for applications than others.In our code, we tried to optimize the algorithm by reducing the number of memories and using them only for the selection and evaluation to minimize the space used [7, 8]. In general, the internal architecture of our algorithm is given by Fig. 1.

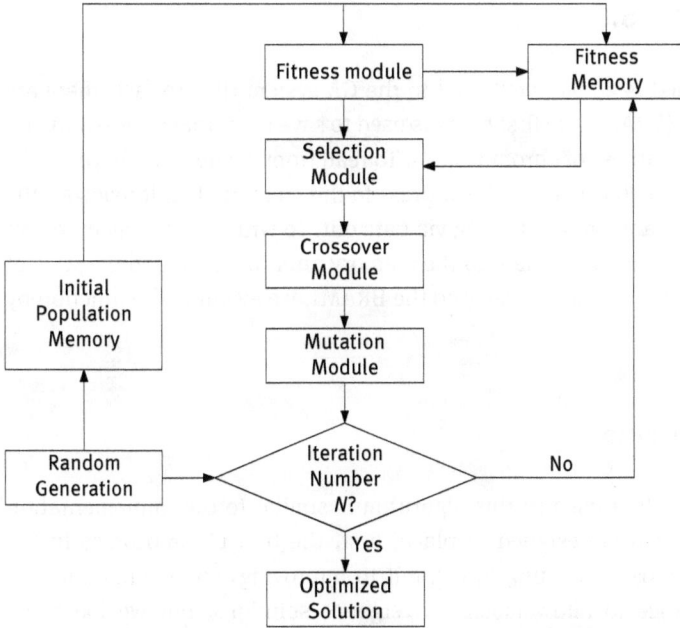

Fig. 1. Internal architecture of the code.

Programming genetic algorithms requires the use of random parameters; there are several methods for generating random numbers.In fact it's impossible to generate a random number based on algorithms and that's why these generators called pseudo-random because the real random exists only in nature. Indeed, it is very difficult to reproduce in a computer except if it is connected to a physical device for measuring a natural phenomenon.

The output of the pseudo-random number is used by two modules. The first supplies a random number generator for the initial population [9] and the second is used in the mutation module. In our application we choose the most commonly used algorithm for generating random numbers. This algorithm is called linear congruence type:

$$X_{n+1} = A.X_n + B \ mod(M) \tag{1}$$

where: X_{n+1} is the $n + 1$ generated value of variable X, X_n is the previous value generated, A is a multiplicative factor, B is a coefficient additive and M is a modulo.

3.1 Initial population

The bloc memory used is actually external to the GA system [10]; in fact, there are two blocs static RAM (BRAM), the first block is used to save the initial population and the second to save the fitness of chromosomes. To read from memory, we have used a finite state machine which transmits the address to the memory then it reads all the chromosomes number and passes it along via data out. To write into the memory, we must transmit address then the data into the memory after that enters to the state of write. After loading the population size into the BRAM1, we evaluate the function by the fitness module.

3.2 Selection module

The GA's selection method used in this algorithm is similar to the implementation of elitism method which is designed to place, first, the best chromosomes in the population who are most promising to commit to improving our population. This method has the chance to allow faster convergence solutions, but we loose the diversity of chromosomes.In fact, we run the risk of excluding people of good quality but could provide good solutions in future generations. The elitism method ensure that the best chromosomes will surely part of the next generation.

3.3 Crossover module

In this section, we will enrich the population by crossing the bad chromosomes. For this we will take part of the solution of certain chromosomes to generate other new chromosomes that will be a good solution to our problem.

3.3.1 Binary example

The new population obtained after selection module is divided into two sub-populations of "N" size and the bad population participates in crossing with a given probability. We can potentially obtain new individuals by crossing the bad population.

If the crossing took place, its location is between position 1 and position "L" in the case of binary encoding, the chromosome is taken according to an uniform law and the two chromosomes share their codes in both sides of this location. Here is an example of crossing located at the third position, which took place between two chromosomes (Fig. 2).

$(12 =)$ | 0 1 1 | 0 0 | 0 1 1 | 1 1 | $(= 15)$

Crossover

$(23 =)$ | 1 0 1 | 1 1 | 1 0 1 | 0 0 | $(= 20)$

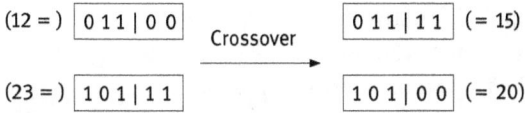

Fig. 2. Example of crossover method.

However, a selected chromosome during reproduction does not necessarily be crossed. The crossing can be made with a fixed probability. When the probability is high then people will support more change.

The Population obtained from the selection module is divided into two chromosomes (Fig. 3) and each pair formed participate in a symmetrical cross (if we have a chromosome coded on 8 bits then the cross point is $\frac{1}{2}N$).

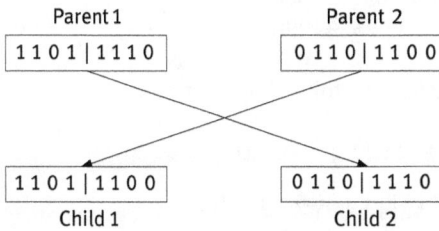

Parent 1   Parent 2
| 1 1 0 1 | 1 1 1 0 | | 0 1 1 0 | 1 1 0 0 |

| 1 1 0 1 | 1 1 0 0 | | 0 1 1 0 | 1 1 1 0 |
Child 1   Child 2

Fig. 3. Symmetric crossing methods.

So if we take "X" parents to cross then we will get "X" new child. We can also use several genes for more "X" new chromosomes. It is not necessary and not recommended to cross all chromosomes in a given population, because nothing tells us if the result of the crossover module will be better or worse than parent's chromosomes.

3.3.2 Real example

There are some difficulties to adapt this encoding because the binary encoding used for genetic algorithms produce some difficulties of optimization of large dimensions and high precision digital problems. The crossing takes place in a simple linear combination between both parents (P_1, P_2) after creating a random number, $\alpha \in [0, 1]$. The new obtained children are:

$$C_1 = \alpha P_1 + (1 - \alpha)P_2 \qquad (2)$$
$$C_1 = (1 - \alpha)P_1 + \alpha P_2 \qquad (3)$$

This operator performs a linear extrapolation of two genes to create new chromosomes that will replace them. After crossover, the chromosomes are subjected to mutation. Mutation prevents the algorithm to be trapped in a local minimum.

3.4 Mutation module

3.4.1 Binary example

Three types of mutation are possible [11]:
–   Deletion of the last bit of the gene
–   Inserting a bit at the end of the gene
–   Modifying a bit in the gene randomly

Mutation module performs a change of one bit randomly and it is applied with a given probability Pm. In random function, we have changed the "prescaler" modulo to not exceed the number of individual encoding.
In the binary code we modify a bit randomly like the following example:

$$\text{Before mutation}: \left(1\,0\,0\,0\,1\,1\,\boxed{0}\,1 \right) \implies (141)_{10}$$
$$\text{After mutation}: \left(1\,0\,0\,0\,1\,1\,\boxed{1}\,1 \right) \implies (143)_{10}$$

3.4.2 Real example

In the real function we have to add an epsilon to the gene, the mutation module allows taking an element of the population with a probability (P_m) and we add a noise "ε_0": such as the following example:

$$e = \varepsilon_0 + \varepsilon \tag{4}$$

with e: mutated element, and ε_0 a fixed noise.

4 Validation examples

In our work we exploit all type of parallelism using a Finite State Machine (FSM) [12, 13], it is a dynamic process, which may have, at every moment a position of several finite numbers of possible states. First, we can define a finite number of states and each transition is around one or more states. States that retain one position and have no transitions are called final states.

To synthesize state machine with standard circuits we must write the equations obtained from the transition diagram. The tool [14, 15] that allows completing transition between diagram and equations is the state and the transition table. So we can define the transition table by following state:
- Input:
  - the presence of a current state of the status register
- Output:
  - presence of input to the next state machine
  - have a future state of the status register
  - outputs equations of the status register must be adapted to the scales used.

We can say that the complexity of the problem increases exponentially as for combinatorial functions [16] if the number of states and the number of inputs increase.

In our algorithm we use the memory modules in pairs to combine the bandwidth and in this way we can maximize the capabilities of the system. It is possible when using Dual Channel RAM. This type of memory allows us to access the contents of the memory in read or write mode in the same frequency but in a different address. The problem of the dual channel RAM that the reading of data is delayed by a clock frequency relative to time of the last reading.

The algorithms proposed in this work have been tested on simple functions in one and two variables. The results are satisfactory and comparable to those obtained by other algorithms developed in the literature. Let's consider the following examples:

$$f(x) = x^2 \tag{5}$$
$$f(x) = x^2 + y^2 \tag{6}$$
$$f(x) = 2x - \frac{1}{2} \tag{7}$$

4.1 Binary encoding examples

4.1.1 Example 1

This section is devoted to the simulation of all operators (selection, crossover and mutation) of the function $f(x) = x^2$ where x is a binary variable coded on 8 bits between 0 and 255.

The contents of memory after selection, crossover and mutation are represented in Fig. 4 and Tab. 1 presents the device utilization of logic table, slice register and the number of input output bloc. The synthesis of Genetic algorithm using a Xilinx Spartan-3 gives results presented in Tab. 1 using Xilinx spartan 3.

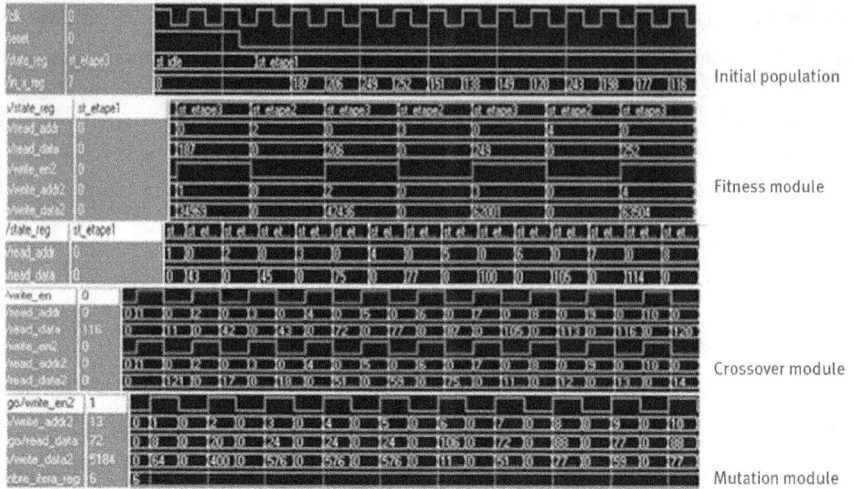

Fig. 4. Simulation result after selection crossover and mutation of example.

Tab. 1. Devices utilizations summary GA Implementation. (^{1}): containing only related logic. (^{2}): containing unrelated logic.

Logic Utilisation	Used	Available	Utilization
Total Number Slice Registers	400	3840	16 %
Number used as Flip Flops	170		
Number used as Latches	230		
Number of 4 input LUTs	629	3840	16 %
Logic Distribution			
Number of occupied Slices	478	1920	24 %
Number of Slices(^{1})	478	478	100 %
Number of Slices(^{2})	0	478	0 %
Total Number 4 input LUTs	640	3840	16 %
Number used as logic	629		
Number used as a route-thru	11		
Number of bonded IOBs	10	173	5 %
IOB Latches	8		
Number of Block RAMs	2	12	16 %
Number of MULT18X18s	1	12	8 %
Number of GCLKs	2	8	25 %

4.1.2 Example 2

Consider the function of two variables: $f(x, y) = x^2 + y^2$. In this case, we will decompose the chromosome into two parts, the 4 most significant bits to the variable "x" and the 4 low significant bits to "y" (Fig. 5).

Note that the first minimum of this function is obtained after the sixth iteration. Simulation results after selection crossover and mutation are represented in Fig. 6 and Tab. 2.

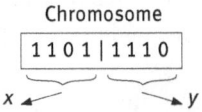

Chromosome

| 1 1 0 1 | 1 1 1 0 |

x $\underbrace{\qquad}$ $\underbrace{\qquad}$ y

Fig. 5. Evaluation function of $f(x, y) = x^2 + y^2$.

Fig. 6. Simulation result after selection crossover and mutation of example 2.

Tab. 2. Devices utilizations summary GA Implementation. (^{1}): containing only related logic. (^{2}): containing unrelated logic.

Logic Utilisation	Used	Available	Utilization
Total Number Slice Registers	408	3840	10 %
Number used as Flip Flops	170		
Number used as Latches	238		
Number of 4 input LUTs	650	3840	16 %
Logic Distribution			
Number of occupied Slices	489	1920	25 %
Number of Slices(^{1})	489	489	100 %
Number of Slices(^{2})	0	489	0 %
Total Number 4 input LUTs	661	3840	17 %
Number used as logic	650		
Number used as a route-thru	11		
Number of bonded IOBs	11	173	6 %
IOB Latches	8		
Number of Block RAMs	2	12	16 %
Number of MULT18X18s	1	12	8 %
Number of GCLKs	3	7	37 %

4.2 Real encoding examples

4.2.1 Example 3

Using the real encoding, the individual becomes a real number in the search space between the interval $[-1, 1]$. The selection operator is identical to that of descending algorithm (selection by elitism). Table 3 summarizes this step.

Function $f(x) = x^2$: the real variable x is coded on 16 bits, and the 16th bit is attributed to the sign. The contents of memory after selection, crossover and mutation are represented in Fig. 7 and the occupied hardware resources are shown in Tab. 4.

We can see easily that the synthesis of a real genetic algorithm using a Xilinx Spartan-3 xc3s200gives an increase of delay time execution and even the using of slice register and Input Output Bloc relative to the binary encoding; the occupation hardware is presented in Tab. 4.

Tab. 3. Contents of memory after selection module.

chromosomes	1	2	3	4	5	6
Initial population	0.4609	0.6093	0.9453	0.9687	0.1796	0.0781
After selection	0.0625	0.0781	0.0937	0.1640	0.1796	0.3828

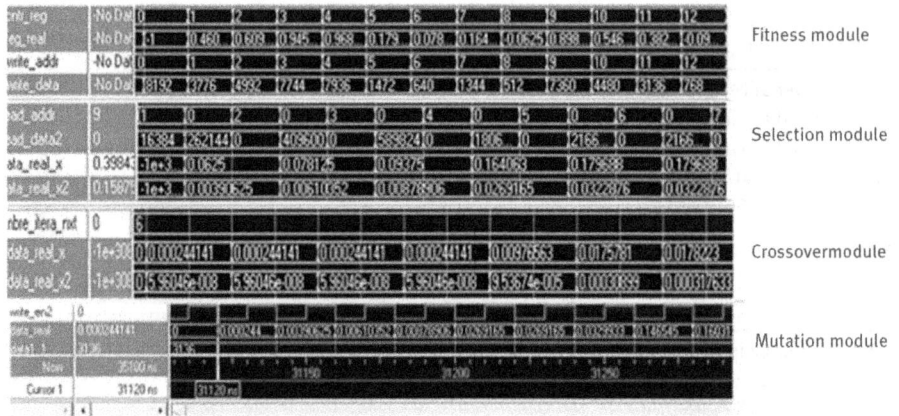

Fig. 7. Simulation result after selection, crossover and mutation of example 3.

Tab. 4. Devices utilizations summary for GA Implementation. (^{1}): containing only related logic. (^{2}): containing unrelated logic.

Logic Utilisation	Used	Available	Utilization
Total Number Slice Registers	603	3840	15 %
Number used as Flip Flops	208		
Number used as Latches	395		
Number of 4 input LUTs	1202	3840	31 %
Logic Distribution			
Number of occupied Slices	847	1920	44 %
Number of Slices(^{1})	847	847	100 %
Number of Slices(^{2})	0	847	0 %
Total Number 4 input LUTs	1271	3840	33 %
Number used as logic	1202		
Number used as a route-thru	69		
Number of bonded IOBs	10	173	5 %
IOB Latches	8		
Number of Block RAMs	2	12	16 %
Number of MULT18X18s	1	12	8 %
Number of GCLKs	5	8	62 %

4.2.2 Example 4

We take the following function $f(x, y) = x^2 + y^2$. In this section we exploit two memories, the first for the chromosome "x" and the second for chromosome "y". The actual code of each chromosome is coded on 15bits and the 16th bit is the sign bit.

The simulation results and the occupied hardware resources adapted to this example are shown in Fig. 8 and Tab. 5.

Fig. 8. Simulation result after selection, crossover and mutation of example 4.

Tab. 5. Devices utilizations summary for GA Implementation of example 4.

Logic Utilisation	Used	Available	Utilization
Number of Slices	1079	1920	56 %
Number of Slice Flip Flops	1011	3840	26 %
Total Number 4 input LUTs	1779	3840	46 %
Number of bonded IOBs	10	173	5 %
Number of Block RAMs	3	12	25 %
Number of MULT18X18s	2	12	16 %
Number of GCLKs	8	8	100 %

4.2.3 Example 5

In this example we change the function and we add a constant real number, like example 3 we used two memories the first for the population coded on 16 bits and the second to fitness chromosomes coded in 32 bits. The function is: $f(x) = 2x - 0.5$. Simulation results and the occupied hardware resources adapted to this example are shown in Tab. 6.

Tab. 6. Devices utilizations summary for GA Implementation. ([1]): containing only related logic. ([2]): containing unrelated logic.

Logic Utilisation	Used	Available	Utilization
Total Number Slice Registers	608	3840	15 %
Number used as Flip Flops	207		
Number used as Latches	401		
Number of 4 input LUTs	1197	3840	31 %
Logic Distribution			
Number of occupied Slices	849	1920	44 %
Number of Slices([1])	849	849	100 %
Number of Slices([2])	0	849	0 %
Total Number 4 input LUTs	1268	3840	33 %
Number used as logic	1197		
Number used as a route-thru	71		
Number of bonded IOBs	10	173	5 %
IOB Latches	8		
Number of Block RAMs	2	12	16 %
Number of GCLKs	5	8	62 %
Total equivalent gate count for design	145174		
Additional JTAG gate count for IOBs	480		

The Spartan-3 Starter board provides a powerful, self-contained development platform for designs targeting the new Spartan-3 FPGA from Xilinx. It features a 200 K gate Spartan-3, on-board I/O devices, and two large memory chips, making it the perfect platform to experiment with any new design, from a simple logic circuit to an embedded processor core. Spartan-3 Starter board has the following features (Fig. 9):

– 200 K gate Xilinx Spartan-3 FPGA with twelve 18-bit multipliers, 216 Kbits of block RAM, and up to 500 MHz internal clock speeds
– On-board 2Mbit Platform Flash (XCF02S)
– 8 slide switches, 4 pushbuttons, 8 LEDs, and 4-digit seven-segment display
– Serial port, VGA port, and PS/2 mouse/keyboard port
– Three 40-pin expansion connectors
– Three high-current voltage regulators (3.3 V, 2.5 V, and 1.2 V)
– Works with JTAG3 programming cable, and P4 & MultiPRO cables from Xilinx

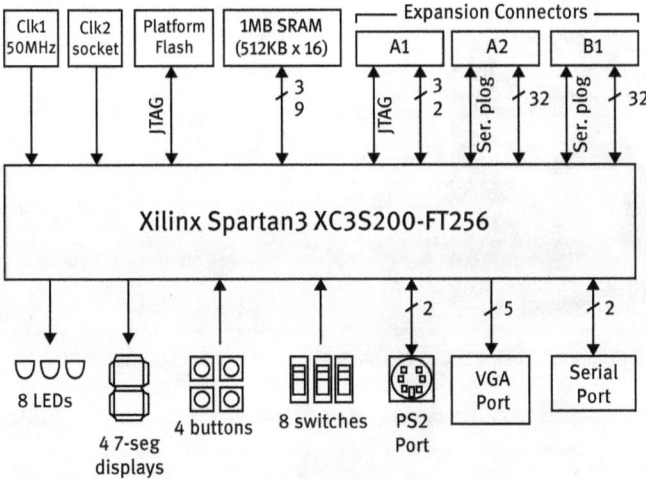

Fig. 9. Block diagram of SPARTAN-3 starter board.

The final implementation in the FPGA board is done with the "Impact" tool of Xilinx allowing loading the "bit stream" into the FPGA memory. The final implementation in kit XC3S200 is done with a specific Digilent cable which is straight compatible with the Xilinx iMPACT software. It loads the bit-stream into the prom of FPGA. Figures 10 and 11 show two photos taken during the practical test of the GA implementation (binary and real example of $f(x) = x^2$) with XC3S200 Spartan3 FPGA. We used two displays modes; the first mode is 4 seven segments module and the second with 8 LEDs at the same time. This allows displaying the iteration number used in the algorithm when our solution tends to "0" and in the second display we use only 8 LEDs.

Fig. 10. Display of the binary function output using the seven segment module and LED module.

Fig. 11. Display of the real function output using the seven segment module and LED module.

5 Conclusion

This work contributes to the implementation on FPGA of some variants of genetic algorithms namely those binary encoding and those to real encoding. For a hardware implementation of the algorithm, FPGAs can exploit all opportunities for parallelism of a given algorithm. By implementing all the concurrent tasks that can be executed simultaneously, the processing time can be reduced in significant proportions.

Thus, in this work, both types have been designed for possible implementations on FPGA while satisfying between the number of used memory and execution time of functions that generate pseudo random codes used in each module (selection, crossover and mutation).

Bibliography

[1]   D.B. Fogel. *Evolutionary Computation*. IEEE Press, 1995.
[2]   I. Rechenberg. *Evolutionary strategie: Evolution*. Frommann-Holzboog, Stuttgart, 1973.
[3]   D.E. Goldberg. *Genetic Algorithms in Search, Optimization and Machine Learning*.
      Addison-Wesley, 1989.
[4]   J. H. Holland. *Adaptation in Natural and Artificial Systems*. University of Michigan Press, 1975.
[5]   I. Lerman and F. Ngouenet. *Algorithmes génétiques séquentiels et parallèles pour une
      représentation affine des proximités*. Research report, INRIA, Rennes, Project REPCO 2570,
      INRIA,1995.
[6]   J. M. Arnold, D. A. Buell and E. G. Davis. Splash-2. 4th *Annual AGM Symposium on Parallel
      Algorithms and Architectures*, :316–324, June 1992.
[7]   Z. Michalewicz. *Genetic Algorithms + Data Structures = Evolutionary programs*. Springer, 3rd
      Edition, 1996.
[8]   W. Stolzmann. *Learning Classifier Systems using the Cognitive Mechanism of Anticipatory
      Behavioral Control. First European Workshop on Cognitive Modelling*, 1996.
[9]   D.H. Lehmer. *Mathematical methods in large-scale computing units*. Ann. Computing Lab.,
      Harvard Univ. :141–146, 1951.
[10] S. Narayanan. Hardware Implementation of Genetic Algorithm Modules for Intelligent Systems.
      IEEE Int. Midwest Symp. on Circuits and Systems, 2005.
[11] G.G. Koonar. *A Reconfigurable Hardware Implementation of Genetic Algorithms for VLSI CAD
      Design*. Thesis, University of Guelph, 2003.
[12] L. Joon-Yong, K. Min-Soeng and L. Ju-Jang. Compact Genetic Algorithms using belief vectors.
      Applied Soft Computing, 11(4):3385–3401, June 2011.
[13] L. Ting and J. Zhu. A genetic algorithm for finding a path subject to two constraints. *Applied
      Soft Computing*, 13(2):891–898, February 2013.
[14] N. Nedjah and L. de Macedo Mourelle. An efficient problem-independent hardware
      implementation of genetic algorithms. *Neurocomputing*, 71(1–3):88–94, December 2007.
[15] H. Fröhlich, A. Košir and B. Zajc. Optimization of FPGA configurations using parallel genetic
      algorithm. *Information Sciences*, 133(3–4):195–219, April 2001.
[16] S. Yussof, R. Azlin Razali and Ong Hang See. A Parallel Genetic Algorithm for Shortest Path
      Routing Problem. *Int. Conf. on Future Computer and Communication*, Kuala Lumpar, Malaysia,
      2009.

Biographies

Mohamed Sadok Ben Ameur was born in Tunisia in 1981. He received the
Engineering degree in electrical engineering and the Master degree in
electronics from the National School of Engineering of Monastir, Tunisia, in
2007 and 2010, respectively. In 2008 he has been with SARTEX Company
where he was the Director of electronic department. Since 2010 he has been
with Tunisian agency of professional training. His current research interests
include rapid prototyping and reconfigurable architecture for real-time
applications.

Anis Sakly is currently a Professor at the Electrical Department, National Engineering School of Monastir (ENIM). He received the Electrical Engineering diploma in 1994 from ENIM, then the PhD degree in Electrical Engineering in 2005 from National Engineering School of Tunis (ENIT). His research interests are in analysis, synthesis and implementation of intelligent control systems. Particularly, his current research interests include soft computing-based approaches applied in optimal control, signal and image processing, and renewable energy systems optimization. He served on the technical program committees for several international conferences.

Abdellatif Mtibaa is currently Professor in Micro-Electronics and Hardware Design with Electrical Department at the National School of Engineering of Monastir and Head of Circuits Systems Reconfigurable-ENIM-Group at Electronic and microelectronic Laboratory. He holds a Diploma in Electrical Engineering in 1985 and received his PhD degree in Electrical Engineering in 2000. His current research interests include System on Programmable Chip, high level synthesis, rapid prototyping and reconfigurable architecture for real-time multimedia applications. Dr. Abdellatif Mtibaa has authored/co-authored over 100 papers in international journals and conferences. He served on the technical program committees for several international conferences. He also served as a co-organizer of several international conferences.

O. Ghorbel, M. W. Jmal, M. Abid, W. Ayedi and H. Snoussi

An Overview of Outlier Detection Technique with Support Vector Machine Developed for Wireless Sensor Networks

Abstract: Wireless Sensor networks (WSNs) is an efficient and emerging area of Computer Science Engineering which has been currently employed in various fields of engineering particularly in communication system to make it effective and reliable. It is important to maintain the basic security level for different types of attacks like both external and internal for successful application of WSNs. Outliers in wireless sensor networks are measurements that deviate from the normal model of sensed data and result from errors, events or malicious attacks on the network. WSNs are more likely to generate outlier due to their special characteristics like constrained available with the resources causing frequent physical failure and harsh deployment area. The dynamic nature of sensor data and the specificity of the wireless sensor network make traditional outlier detection techniques unsuitable for direct application in such contexts so it is essential to select and adapt appropriate techniques to implement in wireless sensor networks for better sensing quality and more reliable system. This paper provides a comprehensive overview of existing outlier detection techniques specifically developed for the wireless sensor networks. Additionally, it presents a technique used to select data type, outlier type and outlier degree. We also investigate applicability of event detection technique for outlier detection. Through experimental study, we evaluate performance of our outlier detection technique to detect outliers and classify them as local or global based on real sensors.

Keywords: Outlier detection, wireless sensor networks, errors, events and malicious attacks.

1 Introduction

A wireless sensor network (WSN) consists of a number of sensor nodes (few tens to thousands) usually deployed in difficult-to-access locations working together to monitor a region to obtain data about the environment. Sensor nodes are small low power devices equipped with one or more sensors, a processor, memory, a power supply, and a radio unit. These units are implemented for wireless communications

O. Ghorbel, M. W. Jmal, M. Abid and W. Ayedi: CES Research Laboratory, National Engineers School of Sfax, University of Sfax, Tunisia, e-mails: oussama.ghorbel@ieee.org, wassim.jmal@gmail.com, mohamed.abid@enis.rnu.tn, ayediwalid@ieee.org.
O. Ghorbel, W. Ayedi and H. Snoussi: University of Technology of Troyes, Troyes, France, e-mails: oussama.ghorbel@ieee.org, ayediwalid@ieee.org, hichem.snoussi@utt.fr.

De Gruyter Oldenbourg, ASSD – Advances in Systems, Signals and Devices, Volume 6, 2018, pp. 145–166.
https://doi.org/10.1515/9783110448375-010

and to transfer the data to a base station where sensor data is analyzed [1]. In any applications of WSNs, real-time data mining of reliable and accurate sensor data is important to make intelligent and fast decisions to maintain the robustness of systems and increase the reliability and safety in critical cases. An appropriate outlier detection technique for the WSN should pay attention to computing, communication and storage limitations of the network and deal with the distributed data analysis. The key objective of outlier detection in WSNs is to identify outliers with a high accuracy while maintaining the resource consumption of the network to a minimum [2].

Many reasons make measured and collected data, by WSNs, unreliable: First, transmitted data streams in the Wireless Network is delicate to errors and sensitive to noise caused by the nature of radio transmissions. Second, the low quality of sensor nodes, including limited computational capacity, restricted energy resources and particularly little communication range, causes sometimes loss of data in the network. The risk to have erroneous, missed or redundant data is high with considering the environmental effects. These effects have a significant impact on the sensor nodes besides the density of the network (up to hundreds or even thousands of nodes) in harsh environments where nodes are vulnerable to malicious attacks i. e. in enemy areas. To improve the quality of sensor data measured by the network, detection of outliers is primordial to ensure reliability and accuracy and to deploy robust, efficient and secure wireless sensor network [3]. The principal problem of outlier detection is due to fraudulent behavior, some changes in system behavior, mechanical faults, instrument error, and human error or simply through natural deviations in populations.

This paper is organized as follows: Brief description of Wireless Sensor Networks was in section 2. In section 3, we present a definition of Outlier, event and their different use in WSNs. Section 4, shows the different classification criteria: types, sources of outlier and then identify outlier data. Section 5, presents a fundamental description of support vector machine (SVM). Section 6; shows the mathematical foundation underlying SVM. Section 7, presents the experimentation and discussion. Conclusion is drawn in section 8.

2 Wireless sensor networks

Wireless Sensor Networks (WSNs) have become a hot topic research in recent years [4]. Their capabilities for monitoring large areas, accessing remote places, real-time reacting, and relative ease of use have brought scientists a whole new horizon of possibilities. So far, WSNs have been employed in military activities such as recognition, surveillance, and target acquisition, in environmental activities [5], or civil engineering such as structural health measurement.

A sensor network typically consists of hundreds, or even thousands, of small, low-cost nodes distributed over a wide area. The nodes are expected to function in an unsupervised fashion even if new nodes are added, or old nodes disappear (e. g., due to power loss or accidental damage). While some networks include a central location for data collection, many other networks operate in an entirely distributed manner, allowing the operators to retrieve aggregated data from any node (Fig 1). Furthermore, data collection may only occur at irregular intervals.

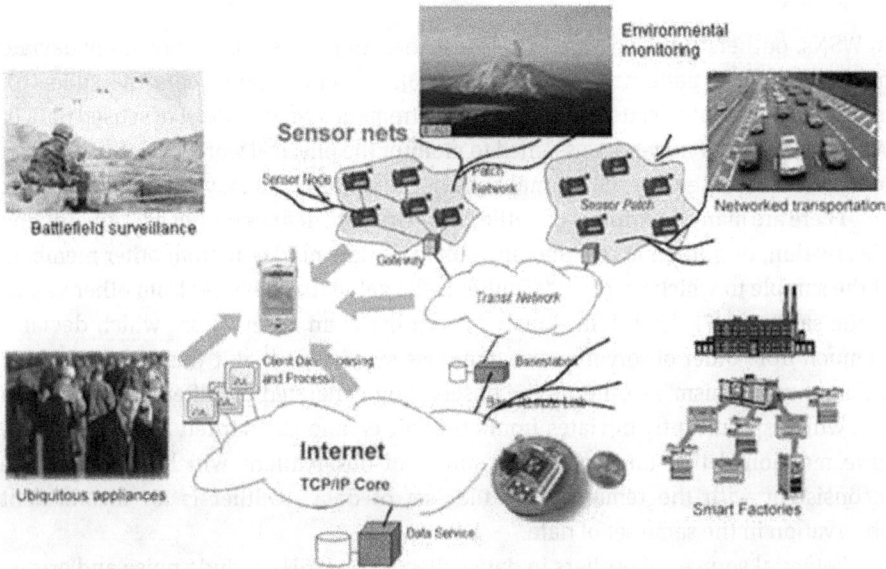

Fig. 1. Architecture of Wireless Sensor Networks.

For example, many military applications strive to avoid any centralized and fixed points of failure. Instead, data is collected by mobile units (e. g., unmanned aerial units, foot soldiers, etc.) that access to the sensor network at unpredictable locations and utilize the first sensor node they encounter as a conduit for the information accumulated by the network [5]. Since these networks often operate in an unsupervised fashion for long periods of time, we would like to detect a node replication attack soon after it occurs. If we wait until the next data collection cycle, the adversary can take advantage of its presence in the network to corrupt data, to decommission legitimate nodes, or otherwise to subvert the network's intended purpose.

3 Outlier detection in wireless sensor networks

We describe a fundamental of Outlier Detection in WSNs, including definitions of outliers, different use of outlier, and events of outlier detection in wireless sensor networks.

3.1 Outlier definition

In WSNs, outliers can be defined as, "those measurements that significantly deviate from the normal pattern of sensed data" [6]. So outlier represents the subset of measurements that deviate in a clear manner from the normal model of sensed data in WSN where sensor nodes are assigned to monitor the physical world and thus a model or a pattern representing the normal behavior of sensed data may exist.

There are many definitions of outlier. So, "Grubbs" defines outlier as: "An outlying observation, or outlier, is one that appears to deviate markedly from other members of the sample in which it occurs". Outlier is the value that deviates from other values of the same set [7]. Then "Hawkins": "An outlier is an observation, which deviates so much from other observations as to arouse suspicions that it was generated by a different mechanism". Outlier is the observation generated by different mechanism and which significantly deviates from other observations. "Barnett and Lewis" also have mentioned that an outlier is a subset of observations which appears to be inconsistent with the remainder of that set of data". Outlier is an inconsistent observation in the same set of data.

Potential sources of outliers in data collected by WSNs include noise and errors, actual events, and malicious attacks. Noisy data as well as erroneous data should be eliminated or corrected if possible as noise is a random error without any real significance that dramatically affects the data analysis [8].

3.2 Use of outlier detection in WSNs

Outlier detection or also called anomaly detection or deviation detection is one of the fundamental tasks in data mining which also includes predictive modeling, cluster analysis and association analysis. Besides, it was subject of research in many domains such as statistics, data mining, machine learning, information theory, and spectral decomposition [9]. It has been also widely applied to numerous application domains such as fraud detection, network intrusion, performance analysis, weather prediction, etc.

Outlier detection is an efficient way to find values that deviate significantly from the other measured data in the network [10]. By the way, these detected values are interpreted as events indicating change of phenomena that are of interest. Besides, outlier detection identifies defected sensors that always generate outlier values, detects potential network attacks, and ensures the security of the network. The value added by outlier detection appears in many real-life applications:

- Habitat monitoring, in which endangered species can be equipped with small non-intrusive sensors to monitor their behavior. Outlier detection can indicate abnormal behaviors of the species and provide a closer observation about behavior of individuals and groups.
- Health and medical monitoring, in which patients are equipped with small sensors on multiple different positions of their body to monitor their well-being. Outlier detection showing unusual records can indicate whether the patient has potential diseases and allow doctors to take effective medical care.
- Industrial monitoring, in which machines are equipped with temperature, pressure, or vibration amplitude sensors to monitor their operation. Outlier detection can quickly identify anomalous readings to indicate possible malfunction or any other abnormality in the machines and allow for their corrections.
- Target tracking, in which sensors are embedded in moving targets to track them in real-time. Outlier detection can filter erroneous information to improve the estimation of the location of targets and also to make tracking more efficiently and accurately.
- Surveillance monitoring, in which multiple sensitive and unobtrusive sensors are deployed in restricted areas. Outlier detection identifying the position of the source of the anomaly can prevent unauthorized access and potential attacks by adversaries in order to enhance the security of these areas.

3.3 Event detection in wireless sensor networks

In literature, WSNs are classified in accordance with four data delivery models: continuous, event-driven, observer initiated and hybrid. In event-driven type, sensors will reply to a request from applications. This model is the most deployed in WSNs applications due to its reactive behavior and the energy expenses of this model are low, when compared to the other cited approaches [11].

Event detection based applications such as military applications for detection of the invasion of enemy forces, health monitoring for detection of abnormal patient behavior, fire detection for setting an alarm if a fire starts somewhere in the monitored area uses techniques to describe events in a way that sensor nodes can understand them using precise values to specify event thresholds.

Event detection techniques includes SQL-like primitives, Petri nets extensions, stochastic methods such Bayesian classifiers and Hidden Markov Models and Fuzzy logic approaches to improve accuracy of event identification. They differ from outlier detections ones for many reasons: they have a priori knowledge of trigger condition or semantic of certain event issued by the sink node [12]. Furthermore, outlier detection techniques aim is to keep the detection rate high and the false alarm rate low by reducing the possibility to classify a normal value as outlier while event detection methods try to exclude erroneous data so that it is not believed as event condition or pattern.

The common issue between event detection and outlier detection is that they both employ spatio-temporal correlations among sensor data of neighboring nodes to distinguish between events and errors and that because sensor faults and noisy measurements are related temporally while events are spatially correlated between each other's.

4 Different classification criteria

Here, we discuss different important aspects and classification criteria of outlier detection techniques developed for Wireless Sensor Networks. We start with differentiation between local models generated from data streams of individual nodes and the global one.

4.1 Types of outlier

The outlier has been divided into two types which are:
- Local Outliers: this type requires that each node identifies the abnormal values only depending on its historical values [8]. Each sensor node collects the transmitted data of its neighboring nodes to identify the anomalous values. Due to the fact that local outliers are identified at individual sensor nodes, techniques for detecting local outliers save communication overhead and enhance the scalability.
- Global Outliers: this is the second type of outlier. We can perform and identify global outliers at different levels in the network [8]. For example, in a centralized architecture, all data is transmitted to the sink node for identifying outliers. This mechanism consumes much communication overhead and delays the response time.

4.2 Identify outlier data

Identifying anomaly data from sensor provides specific methods like Outlier detection techniques. This method can compute the degree of which data measurements deviate from the normal pattern of sensor data. Outliers are measured in two scales, i. e., scalar and outlier score in wireless sensor networks [13].

4.2.1 Scalar

The scalar scale is recognized as a zero-one classification measure, which classifies each data measurement into normal or outlier class. Thus, the output of this technique is a set of outliers and a set of normal measurements. The scalar scale cannot differentiate between different outliers nor provide a ranked list of outliers.

4.2.2 Outlier score

The score scale techniques assign an outlier score to each data measurements. It depends on the degree of which the measurement is considered as an outlier and provides a ranked list of outliers. To select the outliers, an analyst can choose between analyzing top n outliers having the largest outlier scores or using a cut-off threshold. This threshold is usually specified and fixed by the user. In WSNs, the better solution is to learn the threshold and modify it with updates data [14, 15].

4.3 Need of outlier detection in WSN

Outlier is used for finding errors, noise, missing values, inconsistent data, or duplicate data. This abnormal value may affect the quality of data and reduces the system performance. The use of Outlier detection technique is very important in several real life applications, such as, environmental monitoring, health and medical monitoring, industrial monitoring, surveillance monitors and target tracking [16, 17].

In wireless sensor networks, the sensors have low cost and low energy, so to improve the quality and performance, the better solution is to use outlier detection technique.

4.4 Evaluation of outlier detection

Evaluation of an outlier detection technique for WSNs depends on whether it can satisfy the mining accuracy requirements while maintaining the resource consumptions of WSNs to a minimum. Outlier detection techniques are required to maintain a high detection rate while keeping the false alarm rate (number of normal data that are incorrectly considered as outliers) low [18]. A receiver operating characteristic (ROC) curves usually is used to represent the trade-off between the detection rate and false alarm rate.

5 Support vector machine (SVM)

5.1 Fundamentals: hyperplane, margin and support vector

Support vector machine (SVM)-based techniques are from a family of classification-based approaches. The main idea of these techniques is to separate the data belonging to different classes by fitting a hyperplane that produces a maximal margin [24]. SVM possesses several important properties, the most important of which are listed below:

- Maximization of margin. This property can be illustrated by using a linear SVM, which is its simplest form. A linear SVM is a hyperplane that separates the two sets of positive and negative training data by using a maximum margin in the feature space. The margin (m) implies the distance from the hyperplane (i. e. class boundary) to the nearest positive and negative objects in the feature space. The objects closest to the separating hyperplane are called support vectors.
- The distance from the hyperplane to an object approximates the relative strength of the properties that distinguish positive objects from negative ones. For instance, a strong negative object should be located far from the class boundary on the negative side in the SVM feature space.
- Linear or nonlinear transformation of the input space to the feature space can be accomplished by using kernel methods. Advanced kernel methods, such as polynomial and Gaussian kernels, can be used to transform the input space to another high-dimensional feature space when the training data cannot be separated linearly. Linear kernels are fast, but nonlinear kernels have better accuracy for some specific problems.
- The optimal hyperplane is completely defined by using support vectors. That is, a hyperplane built from the entire set of training data is the same as that built from the picked support vectors only. This property is quite helpful where a plentiful supply of training data is difficult to obtain and where support vectors are all that are available.

For two classes of examples given, the goal of SVM is to find a classifier that will separate the data and maximize the distance between these two classes. With SVM, this classifier is a linear classifier called hyperplane.

In the diagram below (Fig. 2), we determine a hyperplane that separates the two sets of points. Then, Fig. 3 shows only the nearest points which are used for the determination of the hyperplane are called support vectors.

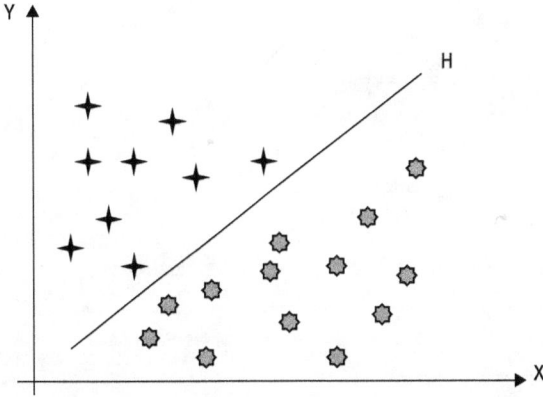

Fig. 2. Hyperplane with two sets of points.

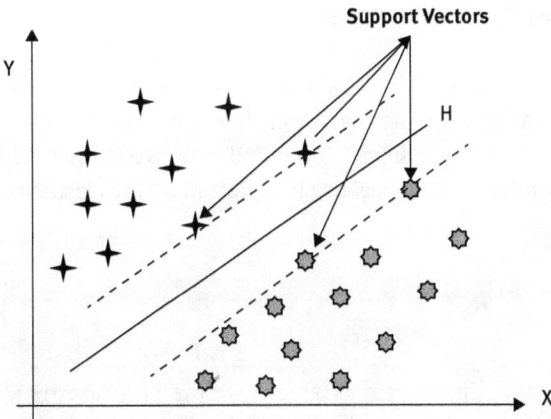

Fig. 3. Support Vectors points.

Obviously there are plenty of valid hyperplane but the remarkable property of SVM is that this hyperplane must be optimal. We will therefore look more valid among hyperplane, one that goes "middle" points of the two classes of examples. Intuitively, this amounts to finding the more precise hyperplane. Indeed, suppose a sample has not been fully described, a small variation will not change its classification if its distance to the hyperplane is great [25]. Therefore, this amounts to finding a hyperplane whose minimum distance to the training examples is maximal. This distance is called "margin" between the hyperplane and the examples. The optimal separating hyperplane is the one that maximizes the margin. As we seek to maximize the margin, this is called vector machines margin.

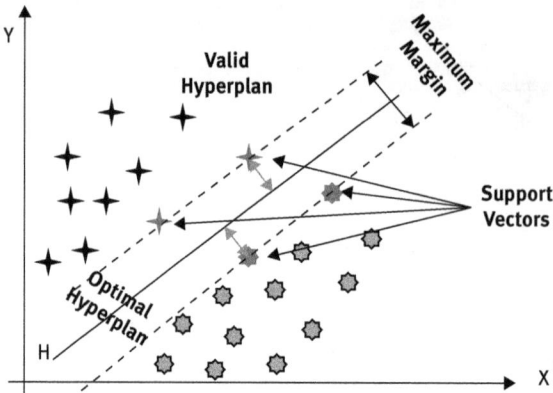

Fig. 4. Vector machine margin.

5.2 Linearity and nonlinearity: linear case

Among the models of SVM, we see the linearly separable and non-linearly separable cases. The first are the simplest SVM because they can easily find the linear classifier [26]. In most real problems there is no linear separation possible between the data, the maximum margin classifier cannot be used because it works only if the classes of training data are linearly separable.

5.3 Nonlinear case

To overcome the disadvantages of non-linearly separable case, the idea of SVM is to change the data space [27]. The nonlinear transformation of the data may allow linear separation examples in a new space. So we will have a change of dimension. This new dimension is called "redescription space". Indeed, intuitively, the higher the dimension of the space of redescription, the greater the probability of finding a separating hyperplane between examples. There is therefore a transformation of a non-linear problem of separation in the space of representation in a problem of separation in a linear space re-description of largest dimension. This nonlinear transformation is done via a kernel function. In practice, few families of kernel functions are configurable and it is known to the user of SVM to perform tests to determine which is best suited for its application [28]. Include the following examples of kernels: polynomial, Gaussian, Laplacian and sigmoid.

6 Mathematical foundations

We will detail in the paragraphs below the mathematical principles underlying SVM.

6.1 Learning problem

We consider a phenomenon f (possibly nondeterministic) which has a certain set of inputs x, produces an output $y = f(x)$. The goal is to find the function f from the single observation of a number of input-output pairs $\{(x_i, y_i), i = 1...n\}$ to "predict" other events.

Knowing h, we can deduce the classification of new points i. e. find a decision boundary. The problem is to find a fairly remote border point of different classes. This is what has been one of the major problems with the classification SVMs.

6.2 Classification with real value

6.2.1 Input transformation

It may be necessary to transform inputs in order to treat them more easily. X is a space of objects. It transforms the input vectors in an F space (feature space) by a function:

$$\varphi : X \longrightarrow F$$

F is not necessarily finite dimensional but has an inner product (Hilbert space). The Hilbert [23] space is a generalization of Euclidean space that can have an infinite number of dimensions. The nonlinearity is treated in this transformation, we can choose a linear separation (we will see later how we manage to bring a nonlinear problem in a classical linear problem) [29, 30]. So, it comes to choosing the optimal hyperplane that correctly classifies data (where possible) and that is as far as possible from all points to classify.

Note that the separating hyperplane chosen should have a maximum margin.

6.2.2 Primal problem

We start with the primal problem and linear variables are introduced to the simplify constraints. A point (x, y) is classified only if $yf(x) > 0$. As the pair (w, b) is defined as a multiplicative factor, it is necessary that: $yf(x) \geq 1$.

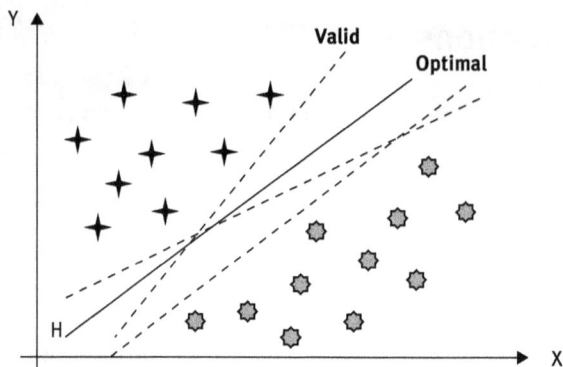

Fig. 5. Vector machine margin.

We deduce (also relying on the previous paragraph), the minimization problem under the following constraints:

$$
\begin{cases}
\min \frac{1}{2}\|w\|^2 + C \sum_{i=1}^{n} \varepsilon_i \\
\forall i, y_i(wx_i + b) \geq 1 - \varepsilon_i
\end{cases}
\tag{1}
$$

$\sum_{i=1}^{n} \varepsilon_i$ relaxes the constraints on the learning vectors, and C is a constant that controls the tradeoff between number of misclassifications and the margin maximization. It may indeed be more comfortable to minimize $\|w\|^2$ rather than directly $\|w\|$.

6.2.3 Dual problem

We go from primal to the dual problem by introducing Lagrange multipliers for each constraint. Here was an example of constraint learning:

$$
\begin{cases}
\max \sum_{i=1}^{n} \alpha_i - \frac{1}{2} \sum_{i,j} \alpha_i \alpha_j y_i y_j x_i x_j \\
\forall i, \alpha_i \geq 0 \\
\frac{1}{2} \sum_{i=1}^{n} \alpha_i y_i = 0
\end{cases}
\tag{2}
$$

This is a quadratic programming problem of n dimension (number of examples). Only those corresponding to the closest points are non-zero. We talk of support vectors. The decision function is associated to:

$$
f(x) = \sum_{i=1}^{n} \alpha_i^* y_i x_i x + b
\tag{3}
$$

However, there are cases where we cannot classify inputs data linearly.

Being penalized by exceeding the constraint, we deduce that the dual problem has the same form as in the separable case:

$$\begin{cases} \max \sum_{i=1}^{n} \alpha_i - \frac{1}{2} \sum_{i,j} \alpha_i \alpha_j y_i y_j x_i x_j \\ \forall i, 0 \le \alpha_i \le C \\ \frac{1}{2} \sum i = 1^n \alpha_i y_i = 0 \end{cases} \tag{4}$$

The only difference is the upper bound C on α.

6.3 Kernel function

In the linear case, we could transform the data into a space or classification would be easier. In this case, the space of redescription is most often used is $\mathbb{R}$ (set of real numbers). It turns out that for the nonlinear case, this space is not sufficient to classify the inputs. Therefore proceed in a high-dimensional space.

$$\begin{array}{cccc} \varphi : & \mathbb{R}^d & \longrightarrow & F \\ & x & \longmapsto & \varphi(x) \end{array} \tag{5}$$

with $\text{card}(F) > d$.

6.3.0.1 Example 1

It is possible to use a linear separation of data, therefore we must solve:

$$\begin{cases} \max \sum_{i=1}^{n} \alpha_i - \frac{1}{2} \sum_{i,j} \alpha_i \alpha_j y_i y_j \varphi(x_i) \varphi(x_j) \\ \forall i, 0 \le \alpha_i \le C \\ \frac{1}{2} \sum i = 1^n \alpha_i y_i = 0 \end{cases} \tag{6}$$

and the solution has the form:

$$f(x) = \sum_{i=1}^{n} \alpha_i^* y_i \varphi(x_i) \varphi(x) + b \tag{7}$$

The problem and its solution depend only on the scalar product $\varphi(x_i)\varphi(x)$. Rather than choosing the non-linear transformation: $\varphi : X \longrightarrow F$, we choose a function $k : X \times X \longrightarrow \mathbb{R}$ (real numbers) called kernel function. It represents an inner product space of intermediate representation. So k is linear (which allows us to reconcile with the linear case of the preceding paragraphs).

This function therefore reflects the distribution of examples in this space: $k(x, x') = \varphi(x) \cdot \varphi(x')$, where k is chosen, we do not need to calculate the representation of examples in this space to calculate φ.

6.3.0.2 Example 2

$$x = (x_1, x_2), \quad \varphi(x) = (x_1^2, \sqrt{2}x_1x_2, x_2^2) \qquad (8)$$

In the intermediate space, the scalar product gives:

$$\varphi(x) \cdot \varphi(x') = x_1^2 x_1'^2 + 2x_1 x_1' x_2 x_2' + x_2^2 x_2'^2$$
$$= (x_1 x_1' + x_2 x_2')^2 = (x \cdot x')^2 \qquad (9)$$

Then, we can calculate $\varphi(x) \cdot \varphi(x')$ without calculating φ: $k(x, x') = (x \cdot x')^2$.

k will thus represent the kernel for the corresponding entries but must nevertheless meet certain conditions.

There are various examples of the kernel function that we present:

- Linear function: $k(x, x') = (x \cdot x')$
- Polynomial function: $k(x, x') = (x \cdot x')^d$ ou bien: $k(x, x') = (C + x \cdot x')^d$
- Gaussian function: $k(x, x') = \exp\left(-\dfrac{\|x - x'\|^2}{\sigma^2}\right)$
- Laplacian function: $k(x, x') = \exp\left(-\dfrac{\|x - x'\|}{\sigma}\right)$

7 Results and discussion

7.1 Experimental dataset

We evaluate our experimentation with two different real data. So, we start with the first scenario which the real data are collected from a closed neighborhood from a WSN deployed as shown in Figure 6. This dataset contains information about data collected from sensors deployed in the lab between November 3rd and 28th, 2012.

Then, the second scenario use real data collected from Intel Berkeley Research lab [31]. This dataset is freely available on their website and contains information about data collected from 54 sensors deployed in the lab between February 28th and April 5th, 2004. This file includes a log of about 2.3 million readings collected from these sensors. The file is 34 MB gzipped, and 150 MB uncompressed. In our work, we have derived column vector of single attributes for individual sensors. These vectors will be used as input dataset for density estimation and outlier detection components. In our experiments, we use a recorded ambient temperature for each sensor measurement.

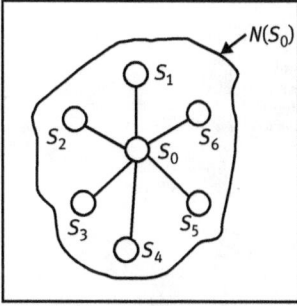

Fig. 6. Deployed WSN with closed sensor nodes.

The closed neighborhood contains the base station S_0 and its 6 spatially neighboring nodes, from S_1 to S_6.

7.2 Results and analysis

For the first scenario, the following figure presents the training set of node S_0 data. It is the case of the panel S_0 which does the apprenticeship phase then it determines the nature of captured data and it classifies it according to the nature of that data. It finds some outliers, normal and support vector data. Figure 8 presents the classified data of nodes S_0 which identifies local outliers. The results are obtained using MATLAB.

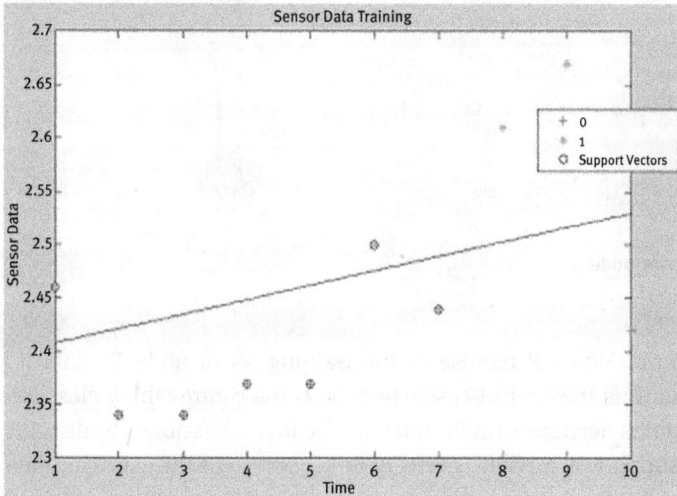

Fig. 7. Training Set of Node S_0.

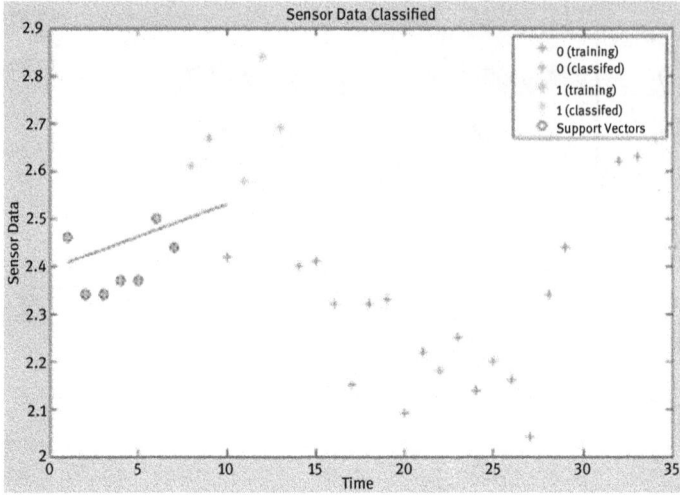

Fig. 8. Local Outliers in the Node S_0.

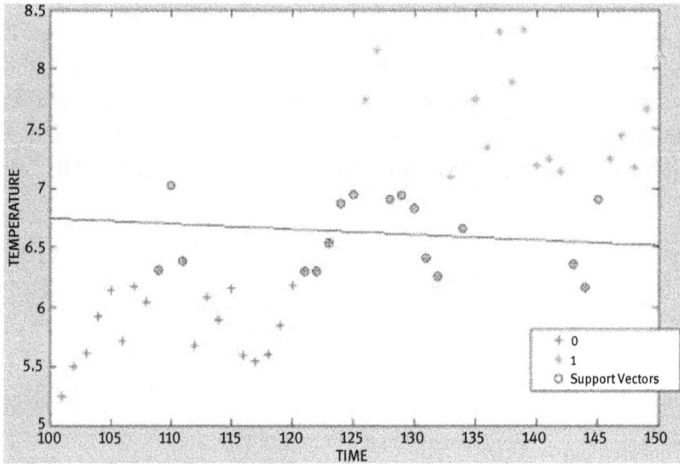

Fig. 9. Training Set of Sensor Node.

For the second scenario, Figure 9 represents the training set of node S_0 data for temperature over time. It is the central case where S_0 is the centre which classifies the data collected from its neighbors (after finishing the first scenario). It finds some outliers, normal and support vector data. Then Figure 10 represents the classified data of node S_0 which identifies local outliers from the other node. The results are obtained using MATLAB.

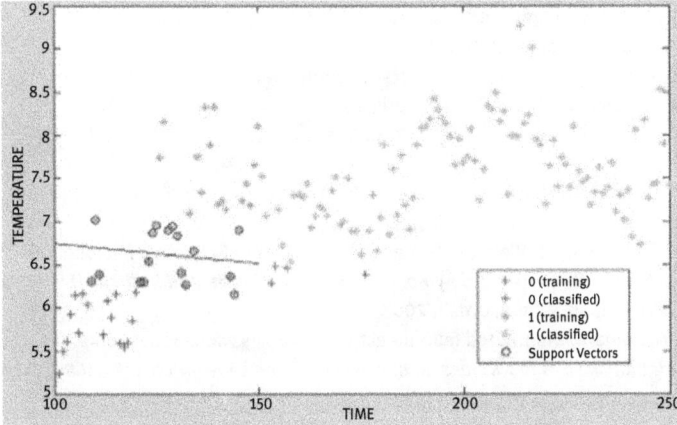

Fig. 10. Classified Sensor Node Data.

8 Conclusion

In this paper we presented outlier detection techniques specifically developed for the wireless sensor networks and we discussed the problem of outliers in this field. We also discussed a classification of outlier detection techniques based on some criteria and a description about WSNs and its characteristics.

It is clearly that there is an insufficiency of techniques for WSNs which calls researchers for developing outlier detection techniques, and takes into account the dependencies of attributes of the sensor node. Researchers also take into account the flexible decision threshold; meet special characteristics of WSNs such as node mobility, and making distinction between errors and events.

Bibliography

[1] I.F. Akyildiz, W. Su, Y. Sankarasubramaniam and E. Cayirci. Wireless sensor networks: a survey. *Computer Networks*, 38(4):393–422, March, 2002.
[2] C.F. Garcia-Hernandez, P.H. Ibarguengoytia-Gonzalez, J. Garcia-Hernandez and J.A. Perez-Diaz. Wireless sensor networks and applications: a survey. *Int. J. of Computer Science and Network Security*, :264–273, 2007.
[3] T. Arampatzis, J. Lygeros and S. Manesis. A survey of applications of wireless sensors and wireless sensor networks. 13rd *Mediterranean Conf. on Control and Automation*, Limassol, Cyprus, :719–724, 2005.
[4] V. Manjula and Dr. C. Chellappan. Replication Attack Miligations for Static and Mobile WSN. *Int. J. of Network Security & Its Applications* (IJNSA), 3(2), March, 2011.
[5] T.G. Lupu. Main Types of Attacks in Wireless Sensor Networks. 9th *WSEAS Int. Conf. on Signal, Speech & Image Processing*, and 9th *WSEAS Int. Conf. on Multimedia, Internet & Video Technologies*, SSIP'09/MIV'09, :180–185, 2009.

[6] Y. Zhang, N. Meratnia and P. Havinga. *Outlier detection Techniques for wireless sensor networks: A survey*. :11–20, Edition !!!, 2008.

[7] C. Zhu, H. Kitagawa, S. Papadimitriou and C. Faloutsos. Outlier detection by example. *J. of Intelligent Information Systems*, 36(2):217–247, April, 2011.

[8] V. Chandola, A. Banerjee and V. Kumar. *Outlier detection: a survey*, Technical Report, University of Minnesota, Country, 2007.

[9] K. Kapitanova, S.H. Son and K. D. Kang. Event Detection in Wireless Sensor Networks. 2nd *Int. Conf. on Ad Hoc Networks* 2010, Victoria, BC, Canada, August 18–20, 2010.

[10] V. Almeida, L. Vieira, B. Vitorino, M. Vieira, A. Fernandes, D. Silva and C. Coelho. Microkernel for Nodes of Wireless Sensor Networks. 3rd *Symp. on Integrated Circuits and Systems Design, Student Forum*, (SBCCI), Chip in Sampa, Brasil, 2003.

[11] J. Chen, S. Kher and A. Somani. Distributed fault detection of wireless sensor networks. *Workshop on dependability issues in wireless ad hoc networks and sensor networks*, :65–72, 2006.

[12] S. Rajasegarar, C. Leckie, M. Palaniswami and J. C. Bezdek. Distributed anomaly detection in wireless sensor networks. *IEEE Int. Conf. on Computational Science*, ICCS, 2006.

[13] T. Joachims. Text Categorization with Support Vector Machines: Learning with Many Relevant Features. *European Conf. on Machine Learning*, Springer, 1998.

[14] F. Markowetz, L. Edler and M. Vingron. Support Vector Machines for Protein Fold Class Prediction. *Biometrical J.*, 45(3):377–389, April 2003.

[15] S. Rajasegarar, C. Leckie, M. Palaniswami and J. C. Bezdek. Quarter sphere based distributed anomaly detection in wireless sensor networks. *IEEE Int. Conf. on Communications*, :3864–3869, 2007.

[16] Y. Zhang, N. Meratnia and P. J. M Havinga. *A taxonomy framework for unsupervised outlier detection techniques for multi-type data sets*. Technical Report, University of Twente, Netherlands, 2007.

[17] Y. Zhang, N. Meratnia and P. J.M. Havinga. Distributed online outlier detection in wireless sensor networks using ellipsoidal support vector machine. *Ad Hoc Networks*, November 2012.

[18] Y. Zhang, N.A.S. Hammb, N. Meratniaa, A. Steinb, M. van de Voorta and P.J.M. Havingaa. Statistics-based outlier detection for wireless sensor networks. 26(8), February, 2012.

[19] V. Barnett and T. Lewis. *Outliers in Statistical Data*. 3rd edition, Hoboken, NJ: Wiley, 1994.

[20] S. Rajasegarar, C. Leckie and M. Palaniswami. Anomaly detection in wireless sensor networks. *IEEE Wireless Commununications*, 15(4):34–40, August, 2008.

[21] S. Rajasegarar, C. Leckie and M. Palaniswami. Detecting data anomalies in wireless sensor networks. *Security in Ad-Hoc and Sensor Networks*, R. Beyah, J. McNair and C. Corbett, Editors, Singapore: World Scientific, :231–260, July 2009.

[22] V. Chandola, A. Banerjee and V. Kumar. Anomaly detection: A Survey. *ACM Computing Surveys* 41(3):1–58, July 2009.

[23] R. Pincus, V. Barnett and T. Lewis. *Outliers in Statistical Data*. 3rd edition, J. Wiley & Sons 1994, XVII. 582, :49–95, *Biometrical Journal*, 37(2):256–xxx, 1995.

[24] D. M. Tax and R. P. Duin. Support Vector Data Description. Machine Learning, 27(4):45–66, August, 2004 .

[25] D. M. Tax and R. Duin. Outliers and data descriptions. 7th *Annual Conf. of the Advanced School for Computing and Imaging*. :234–241, Citeseer, 2001.

[26] E. Flouri, B. B. Lozano and P. Tsakalides. Training A SVM-Based Classifier In Distributed Sensor Networks. 14th *European Signal Processing Conf.*, September, 2006.

[27] H. Yu, X. Jiang and J. Vaidya. Privacy-Preserving SVM Using Nonlinear Kernels On Horizontally Partitioned Data. *ACM SAC Data Mining Track*, :603–610, April 2006.

[28] C. Domeniconi and D. Gunopoulos. Incremental support vector machine construction. *IEEE Int. Conf. on Data Mining*, ICDM'01, :589–592, November, 2001.

[29] K. Flouri, B. B. Lozano and P. Tsakalides. Distributed Consensus Algorithms for SVM Training in Wireless Sensor Networks. 16th *European Signal Processing Conf.*, Lausanne, Switzerland, 2008.

[30] T. J. Dodd, V. Kadirkamanathan and R.F. Harrison. Function estimation in Hilbert space using sequential projections. *Conf. on Intelligent Control Systems and Signal Processing*, :113–118, 2003.

[31] http://sensorscope.epfl.ch/index.php/MainPage.

Biographies

Oussama Ghorbel is a Ph.D student at the National Engineering School of Sfax since January 2011. Her research activity is conducted within CES research unit. He has received the Diploma degree in Computer Science, from the Faculty of Sciences of Sfax, Tunisia in 2007, the Engineering degree from the National Engineering School of Sfax, in 2009 and the Master degree in New Technologies of Dedicated Computer Science Systems, from the National Engineering School of Sfax, in 2010. Her current research interests are in the field of Wireless Sensor Networks (WSN) and Image Compression. He served in national and international conference organization: ICM, TWESD, and SensorNets-09.

Mohamed Wassim Jmal obtained PhD degree on Electrical Engineering at the National Engineering School of Sfax, Tunisia since 2013. His research activity is conducted within CES Laboratory. He has received the Engineering degree in Electrical Engineering, from the National Engineering School of Sfax in 2005 and the Master degree in Automatic and Industrial Informatics, from the same Engineering School, in 2007. His current research interests are in the field of Wireless Sensor Networks (WSN) and the Embedded Systems. They are focused on the implementation of wireless sensor networks applications in Reconfigurable System He is working now as Assistant in Higher Institute of Applied Science and Technology of Gafsa, Tunisia. Mohamed Wassim JMAL has served in national and international conference organization: IDT, ICM, TWESD, and Sensor Nets.

Mohamed Abid Head of "Computer Embedded System" laboratory CES-ENIS, Tunisia. Mohamed ABID is working now as a Professor at the Engineering National School of Sfax (ENIS), University of Sfax, Tunisia. He received the Ph. D. degree from the National Institute of Applied Sciences, Toulouse (France) in 1989 and the "thèse d'état" degree from the National School of Engineering of Tunis (Tunisia) in 2000 in the area of Computer Engineering & Microelectronics. His current research interests include: hardware-software co-design, System on Chip, Reconfigurable System, and Embedded System, etc. He has also been investigating the design and implementation issues of FPGA embedded systems. He was founding member and responsible of doctoral degree "computer system engineering" at ENIS, 2003–2010. Dr. Abid served in national and international conference organization and program committees at different organizational levels. He was also Joint Editor of Specific Issues in two International Journals. Dr. Abid is joint coordinator or an active member of several International Re-search and Innovation projects: STIC/INRIA project since 2009, CMCU project and since 2009 and Head of Federator Research Project since 2009. Dr. Abid was Supervisor or Co-supervisor of more than 20 PhD doctors, most of them were in joint guardianship and Supervisor or Co-supervisor of more than 50 master students. He is Author or co-author of more than 30 publications in Journals and author or co-author of more than 180 papers in international con-ferences. He is also author or co- author of many guest's papers, Joint author of many book's chapters. Dr. Abid has served also as Guest professor at several international universities and as a Consultant to research & development in Telnet Incorporation.

Walid Ayedi received the MS Degree from National Engineering School of Sfax, Tunisia in 2008. He is a PhD student at Computer & Embedded Systems Laboratory, Tunisia. He is actually an invited PhD student in University of Technology of Troyes, France, at the Laboratory of Systems Modeling and Dependability. His research interests include image analysis, computer vision and machine learning. He served in national and international conference organization: IDT, ICM, TWESD, and Sensor Nets.

Hichem Snoussi was born in Bizerta, Tunisia, in 1976. He received the diploma degree in electrical engineering from the Ecole Superieure d'Electricite (Supélec), Gif-sur-Yvette, France, in 2000. He also received the DEA degree and the Ph.D. in signal processing from the University of Paris-Sud, Orsay, France, in 2000 and 2003 respectively. Between 2003 and 2004, he was postdoctoral researcher at IRCCyN, Institut de Recherches en Communications et Cybernétiques de Nantes. He has spent short periods as visiting scientist at the Brain Science Institute, RIKEN, Japan and Olin Neuropsychiatry Research Center at the Institute of Living in USA. Between 2005 and 2009, he was associate professor at the University of Technology of Troyes, France. He has obtained the HDR degree from the University of Technology of Compiègne in 2009. Since 2010, he is Full Professor at the University of Technology of Troyes. His research interests include Bayesian techniques for source separation, information geometry, differential geometry, machine learning, robust statistics, with application to brain signal processing, astrophysics, advanced collaborative signal/image processing techniques in wireless sensor/cameras networks, Since January 2011, he is leading the research group "Risk Management of Complex Systems and Networks" of the CNRS STMR UMR Laboratory. He is in charge of the regional research program S3 (System Security and Safety) of the CPER 2007–2013 and the CapSec platform (wireless embedded sensors for security). He is the principal investigator of an ANR-Blanc project (mv-EMD), a CRCA project (new partnership and new technologies) and a GDR-ISIS young researcher project. He is partner of many ANR projects, GIS and strategic UTT programs. In 2009, he launched a new company Track & Catch on smart embedded cameras for security and surveillance, where he is the scientific director. He is author of more than 100 research papers in journal and international conferences. He is member of organizing and program committee of many conferences. He obtained the national doctoral and research supervising award PEDR 2008–2012.

S. Bedoui, H. Charfeddine Samet and A. Kachouri

Electronic Nose System and Principal Component Analysis Technique for Gases Identification

Abstract: An electronic nose is an intelligent system which consists of a sensor network and a pattern recognition system able to know simple and complex odors. As the human nose, the artificial nose must learn to recognize different odors: the learning phase. There are several types of sensors such as fiber optic sensors, piezoelectric sensors, and sensor type MOSFET. The performance of the sensor network is discussed by using pattern recognition methods. In this paper, we tested Principal Component Analysis (PCA) to evaluate the ability of our sensor array to distinguish between different groups of target gases according to their nature: only in binary mixture and ternary mixture.

Keywords: Electronic nose, sensors array, gases, PCA.

1 Introduction

PCA is a simple method used to project data from several sensors to a two-dimensional plane [1].

Principal component analysis was heavily used in various research aimed to classify different types of data. This method has been applied for the analysis of non-stationary variables [2], and to defect detection with applications to pollution parameters of the region of Annaba [3]. This technique has been chosen to detect changes of the land [4].

Another application of this method was for assessing the capabilities of instrumental techniques for discriminating marine oils and studying the positional distribution of fatty acids on the backbone of triacylglycerols [5]. Unsupervised pattern recognition techniques like Principal Component Analysis (PCA) have widely been used for odor identification [6]. Wu and Kuo in [7] used air quality data collected from eight automatic air quality monitoring stations in central Taiwan and discussed the correlation between air quality variables with statistical analysis in an attempt to accurately reflect the difference of air quality observed by each monitoring station as well as to establish an air quality classification system suitable for the whole Taiwan. The overall objective of [8] is to propose a novel methodology for identifying and

S. Bedoui, H. Charfeddine Samet and A. Kachouri: Department of Electrical Engineering, LETI Laboratory, , National School of Engineers of Sfax, University of Sfax, Tunisia. , Emails: souhir.bedoui@yahoo.fr, hekmet.samet@enis.rnu.tn, abdennaceur.kachouri@enis.rnu.tn.

De Gruyter Oldenbourg, ASSD – Advances in Systems, Signals and Devices, Volume 6, 2018, pp. 167–180.
https://doi.org/10.1515/9783110448375-011

apportioning air pollution sources in a French urban site. The identification of the sources profiles was achieved through Principal Component Analysis. PCA was also applied in [9] to compare air pollutions profiles of cities.

In this paper, we test the PCA for the detection of gases atmospheric pollutants. In fact, due to various human activities, air pollution is increasing more and more. In this context, several studies have aimed to reduce this pollution. In addition, this pollution often has unpleasant odor, hence the need to control the atmosphere. This control requires very expensive and very complex techniques, hence the idea of designing an intelligent "artificial nose".

An electronic nose is a device used for the detection of gaseous mixtures. It is the subject of several research studies, especially in Europe and the United States for different applications such as medicine, pharmaceutical, environmental, etc...

Electronic nose consists of an array of chemical sensors and an electronic system capable of recognizing simple and complex odor patterns. The electronic nose is an instrument able to discriminate and identify odors. The principle of an electronic nose is inspired by the human nose. It consists of an array of sensors sensitive to odor components, a system for conditioning the signal from the network and data processing software.

Several supervised methods determine decision rules, which allow the identification of unknown gas samples, which is the purpose of the electronic nose. A sufficient number of measurements is made to form a database for the learning phase. Thus coefficients extracted will be implemented as an algorithm in the intelligent part.

There are several types of sensors such as fiber optic sensors, piezoelectric sensors, and sensor type MOSFET. Several steps must be applied to the sensor network; each measurement should be repeated more than 10 times, taking into account all possibilities (humidity, gas temperature, concentration ...) to fully characterize the response of a set of sensors. Performances of the sensor network are discussed by using pattern recognition methods. Figure 1 presents our approach to identify a gas.

There are several classification methods such as Principal Component Analysis which is the subject of this article. This paper is organized as follows. In sections 2

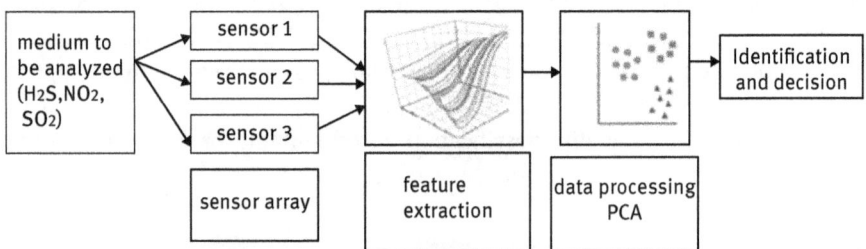

Fig. 1. Proposed approach to identify a gas.

and 3, we present the sensors array, the type of sensors used in our application and the characteristic parameters. In section 4, we briefly describe the principal component analysis method. The application of this method and the results are described in section 5 and finally our conclusion is given in section 6.

2 Sensors array

Metal oxide sensors, piezoelectric quartz sensors and polymers sensors are the most widely used for electronic nose.

Metal Oxide Semiconductor (MOX) gas transducers are one of the most preferable technologies to build electronic noses thanks to their high sensitivity and low price [10]. The sensor network is composed of three types of gas sensors Taguchi Gas Sensor (TGS.) type 8xx and 2xxx sensitive to different atmospheric gases.

Unfortunately, as mentioned in [11], the ambient humidity is an important factor affecting the performance of metal oxide gas sensor because the water absorption reduces the sensitivity of these sensors.

The Electronic nose system used for the gas detection depends on the resistance variation of the gas sensor, for possible increases of the selectivity and sensitivity [12]. Table 1 summarizes the sensors used.

Tab. 1. Recapitulative of used sensors.

	resistance	Target gas
TGS826	46 kΩ	NH$_3$
TGS2106	3.9 kΩ	NO$_2$
TGS2610	33 kΩ	Combustion gases

3 Feature extraction

First, we characterize the sensors in the presence of a gas insulated, in a binary mixture, and a ternary mixture.

H$_2$S is studied in concentrations close to emissive sources (1, 4 and 7 ppm).

We studied the response of the sensor network to NO$_2$ (1, 3 and 5 ppm).

The levels of SO$_2$ in the atmosphere are higher than those of H$_2$S and NO$_2$. For that, we studied the response of sensors in the presence of SO$_2$ (5, 10, 15 and 20 ppm). In case of binary mixture, the three gases H$_2$S, NO$_2$ and SO$_2$ may be combined into three binary groups.

For example, for the mixture (H_2S/NO_2), the concentration of one gas is fixed and the concentrations of the other gas are variable then the measurements were repeated by reversing the two gases as shown in Tab. 2

Tab. 2. Protocol change in concentration of H_2S and NO_2.

Step	NO_2 (ppm)	H_2S (ppm)		Step	NO_2 (ppm)	H_2S (ppm)
1	1	1		1	1	1
2	1	4		2	3	1
3	1	7		3	5	1
4	3	1	Then	4	1	4
5	3	4		5	3	4
6	3	7		6	5	4
7	5	1		7	1	7
8	5	4		8	3	7
9	5	7		9	5	7

The evolution of the gas during 10 minutes has been presented in Fig. 2 and Fig. 3.

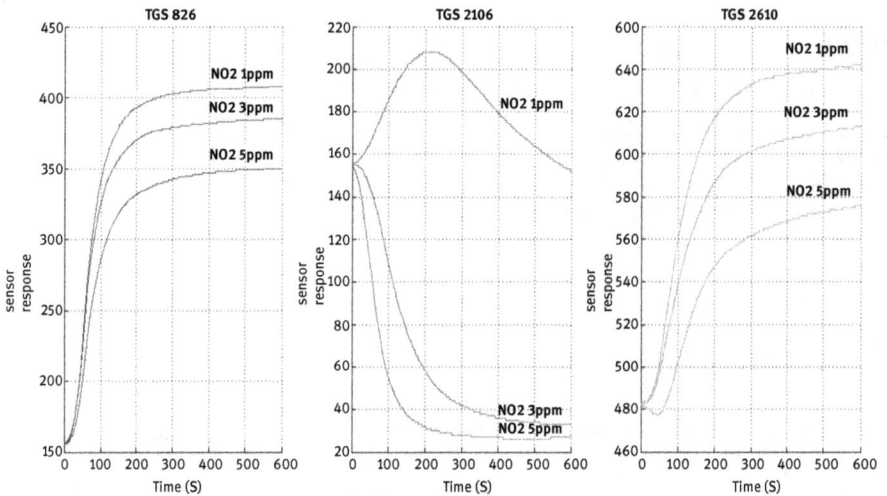

Fig. 2. Curves of the mixture of H_2S gas (4 ppm) and NO_2 gas at a variable concentration from the three sensors.

The characteristic parameters of sensors are the initial conductance G_0, the final conductance G_s, G_m the gap between the maximum and minimum conductance

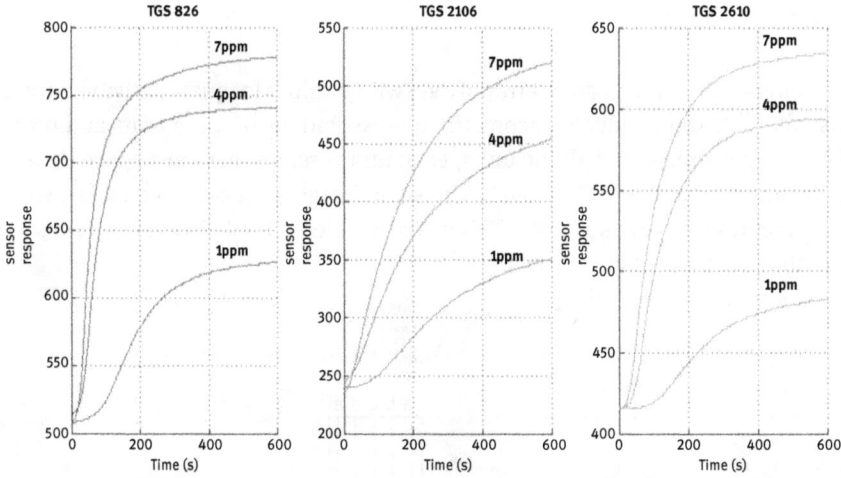

Fig. 3. Curves of H_2S from the three sensors.

and the sensitivity S. In this paper, we focus on the implementation of the PCA. All simulations were performed using MATLAB environment. We applied this method on the variable S.

$$G_0 = \frac{1}{20} \sum_{k=1}^{20} G_k \qquad (1)$$

$$G_s = \frac{1}{20} \sum_{k=580}^{600} G_k \qquad (2)$$

The sensitivity is defined as follows:

$$S = \frac{G_s - G_0}{G_0} \qquad (3)$$

The sensitivity advantage is that it reduces the conductance values, which vary widely from one sensor to another, in the same order.

4 Principal component analysis (PCA)

The PCA method is a technique used to reduce the number of space dimensional of parameters or characteristics presented to the input of a system for classification or different treatment. PCA is a mathematical procedure that uses an orthogonal transformation to convert a set of observations of possibly correlated variables into a set of values of linearly uncorrelated variables called principal components [13].

4.1 Table of data

The PCA offers, from a rectangular array of data with p values for quantitative variables n units (also called individuals), geometric representations of these units and these variables. To get a picture of all the units, each unit is represented in a space R_p this image is called "cloud of individuals". The same image for a set of variables, each variable is represented in a space Rn. This image is called "cloud of variables" (Fig. 4).

- Situation A: cloud of individuals.
- Situation B: cloud of variables.

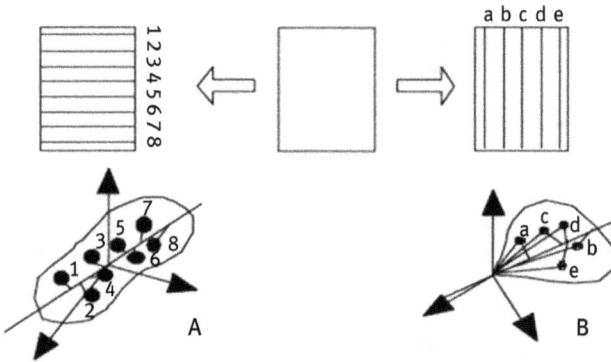

Fig. 4. The cloud of variables and the one of units.

4.2 Search principal components

Let's note $C_1, C_2, ..., C_k, ..., C_q$ the main components having a linear combinations of original variables $X_1, ..., X_p$:

$$C_k = a_{1k}X_1 + a_{2k}X_2 + ...a_{pk}X_p = \sum_{i=1}^{p} a_{ik}X_i \qquad (4)$$

Coefficients a_{jk} are determined such that all C_k are 2–2 uncorrelated variance and maximum of decreasing importance. The first principal component C_1 must be of maximum variance.

Figure 5 presents the first principal component C_1 when C_{i1} is the coordinate of point i on the axis C_1 with:

$$C_{i1} = \sum_{j=1}^{p} a_{1j}X_{ij} \qquad (5)$$

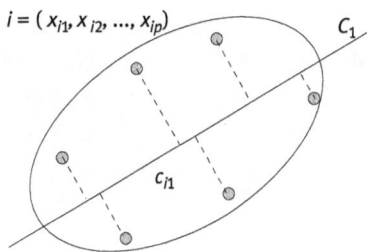

Fig. 5. Principal components.

This line passes through the center of gravity and ensures minimal distortion. C_2 is the second principal component. This latter is orthogonal to C_1 and of maximum variance.

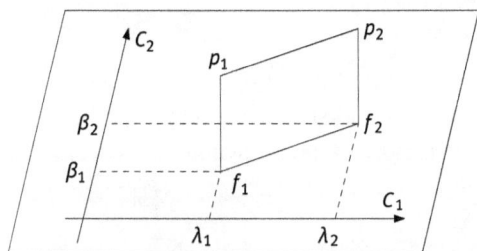

Fig. 6. Projection plane.

It is clear from Fig. 6 that C_1 and C_2 determine the best plan with projection:
- For the average C_1, $d_2(\lambda_1, \lambda_2)$ is maximum;
- For C_2 (C_1 perpendicular to) the average of $d_2(\beta_1, \beta_2)$ is maximal.

Similarly, we determine the other major components. It is interesting to see how variables are related to old and new ones. For this, we calculate correlations of the old ones with the new variables.

Tab. 3. Correlations of the old with the new variables.

	C_1	C_2	C_3	$\cdots$
X_1	r_{11}	r_{12}	r_{13}	$\cdots$
X_2	r_{21}	r_{22}	r_{23}	$\cdots$
$\vdots$	$\vdots$	$\vdots$	$\vdots$	$\vdots$
X_p	r_{p1}	r_{p2}	r_{p3}	$\cdots$

Then, we present the old variables taking into account their correlation coefficients with the news. These coefficients are the coordinates of the old variables. We then obtain the correlation circle (Fig. 7). Background variables are presented as points within a circle of radius 1.

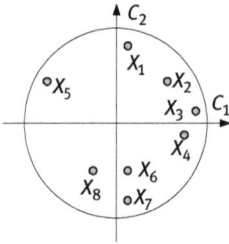

Fig. 7. Circle of correlation.

From this circle of correlation, we can study the link between the variables. For example, two points that are very close to the circle are positively correlated, and two points very close to the circle, in relation to the origin having symmetrical positions, are negatively correlated.

5 Identifying groups of gaseous mixtures by PCA

The sensitivities of 3 sensors for each target gas were used as input parameters for PCA.

5.1 Identification of three groups of gas mixture

First, we study the separation into three group gaseous alone, binary mixture and ternary mixture. Figure 8 shows a good discrimination between the three groups.

After performing the PCA, we obtain two axes. In fact, these two axes correspond to the highest eigenvalues and they contain most of the information. These two axes are called principal components.

The principal components are chosen to contain the maximum data variance and to be orthogonal. The percentage of the data variance contained in each principal component is given by the corresponding eigenvalue. In fact, the variance gives us an idea of the dispersion of information. The first component must be of maximum variance. PCA shows that each of the three samples groups was discriminated from the each other The first principal component expresses 99.04 % of the total variance in data.

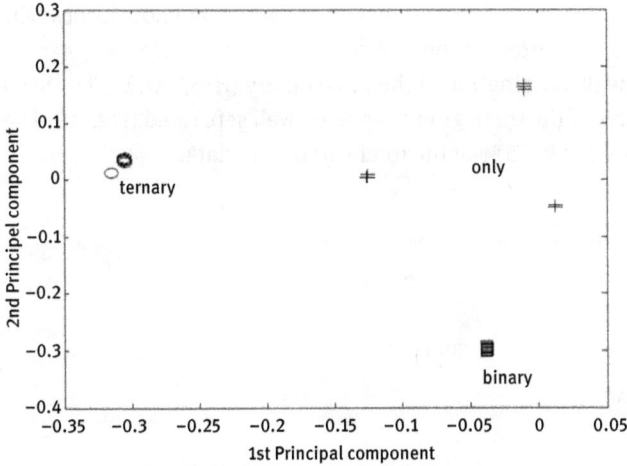

Fig. 8. Identification of three groups of gaseous mixtures.

5.2 Group "gas alone"

To separate the three gases, we applied the same method for the "gas alone". Figure 9 shows that a linear method such PCA is able to correctly separate the clusters associated with NO_2, SO_2 and H_2S.

The first principal component accounted for 99.1 % of the total variance in data. From Fig. 9, we see that the distribution of poles for NO_2 is wide. This is caused by

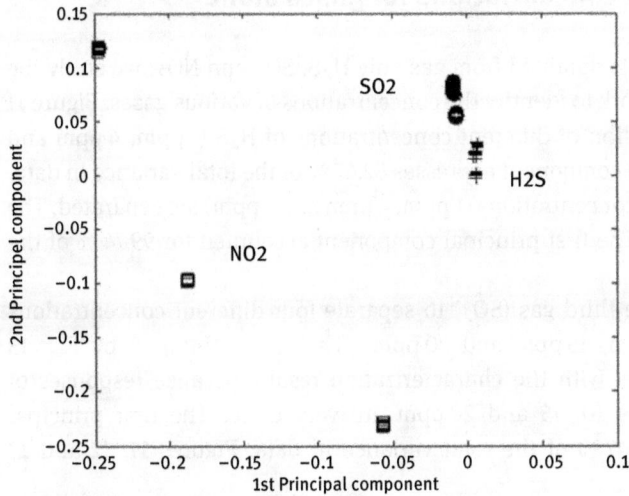

Fig. 9. Group "gas alone".

answers given by the three sensors in the presence of gas which are not closely. SO_2 and H_2S samples were closely located on the PCA plot

The analysis allows the discrimination of the three binary (H_2S/NO_2), (SO_2/NO_2) and (SO_2/H_2S). The analysis of the three groups appears well separated (Fig. 10). The principal component expressed 92.75 % of the total variance in data.

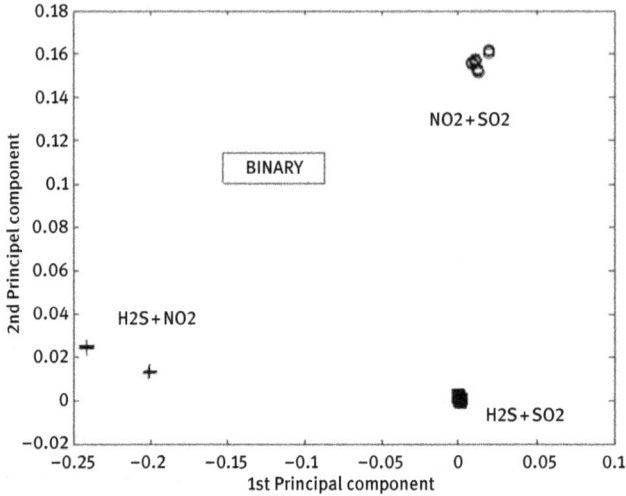

Fig. 10. Binary mixture.

5.3 Identification of concentrations for gases alone

Using measurement results obtained from gas only H_2S, SO_2 and NO_2, we study the ability of our sensor network to identify the concentrations of various gases. Figure 11 shows a good discrimination of different concentrations of H_2S (1 ppm, 4 ppm and 7 ppm). The first principal component expresses 82.67 % of the total variance in data.

For NO_2 also various concentrations (1 ppm, 4 ppm and 7 ppm) are separated. The result is given in Fig. 12. The first principal component accounted for 99.74 % of the total variance in data.

PCA is applied to the third gas (SO_2) to separate four different concentrations of the gas (5 ppm, 10 ppm, 15 ppm and 20 ppm). The result obtained by Fig. 13 presents good agreements with the characterization results because responses of sensors to concentrations 10, 15 and 20 ppm are very close. The first principal component expresses 99.47 % of the total variance in data. Figures 11, 12 and 13

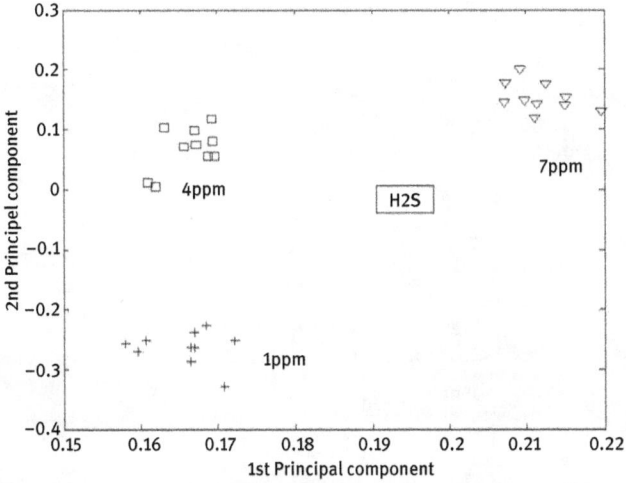

Fig. 11. Identification of gas concentrations for H_2S.

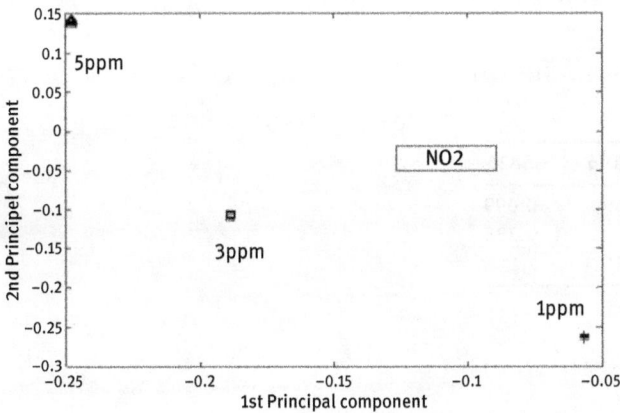

Fig. 12. Identification of gas concentrations for NO_2.

show the good ability of our sensor network. Concentrations are identified and grouped according to their value. In fact PCA shows that samples which have the same concentration were discriminated from others. In order to replace redundant sensors by other sensors which may be more specific, we decided to determine the correlation between the three sensors. If we consider the sensors as variables, we can evaluate their interdependence with correlation matrix (Tab. 4). The strongest correlation is 73 %, so the sensors are not too correlated, which confirms our right choice.

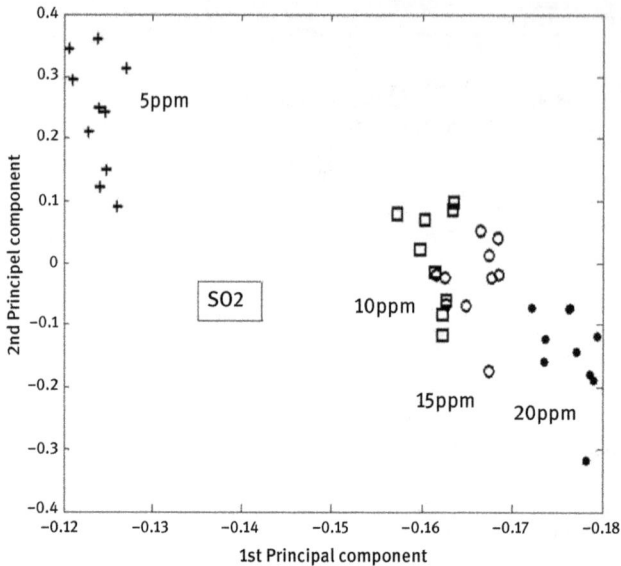

Fig. 13. Identification of gas concentrations for SO_2.

Tab. 4. Correlation matrix.

	TGS2610	TGS2106	TGS826
TGS2610	1	0.259	0.099
TGS2106	0.259	1	0.738
TGS826	0.099	0.738	1

6 Conclusion

In this paper, we presented a method for pattern recognition: the PCA. This method gave good results and showed good ability of our sensor network to discriminate between different target gases H_2S, NO_2 and SO_2.

At first, we tried to separate these three gas mixtures (alone, binary and ternary). Then we refined the analysis by seeking gases in groups of gaseous mixtures (alone or binary) . Finally, we tried to identify the concentrations of gas for each gas alone.

PCA provides a classification according to the type of gas that can be "alone" in "binary mixture" or "ternary". In fact, this method allows using linear transformations passing a large set of correlated data to a smaller set.

Acknowledgment: This work has been supported by CMPTM 13/TM32. Authors would like to acknowledge the Tunisian-Moroccan cooperation in scientific research and technology, specially LETI and OSCARS laboratories.

Bibliography

[1] H.-K. Hong, C. H. Kwon, S.-R. Kim, D. H. Yun, K. Lee and Y. K. Sung. *Portable electronic nose system with gas sensor array and artificial neural network.* Devices and Materials Laboratory (MS: MA Gr.), LG Corporate Institute of Technology, Seoul, South Korea, 1999.

[2] P. Casin, C. Stachowiak and F. Marque. L'analyse en composantes principales de variables non stationnaires. *Mathematics & Social Sciences*, 49(4):27–40, 2011.

[3] B. Ouassim, M. T. Khadiry and D. Messaoudz. *Analyse en Composante Principale Multi-Echelle Non Linéaire pour la détection de défauts avec applications aux paramètres de pollution de la region d'Annaba.* Département Electronique, Université Badji-Mokhtar Annaba, Algérie, 2009.

[4] K. H. Hoang, M. Bernier and J.-P. Villeneuve. Les changements de l'occupation du sol dans le bassin versant de la rivière cau (Vietnam), Essai sur une approche diachronique. *Revue Télédétection*, 8(4):227–236, 2008

[5] Y. Zeng. *Exploring the capabilities of gas chromatography and liquid chromatography single and tandem mass spectrometry for discriminating and characterizing marine oils by using chemometric tools.* Master Thesis, National Institute of Nutrition and Seafood Research, Bergen, Norway, February 2010.

[6] A. Agrawal, R. Kumar and A. K. Kohli. *Performance Comparison of PCA and ICA Pre-processors in Identification of Individual Gases Using Response of a Poorly Selective Solid-state Sensor Array.* Electronics and Communication Engineering Department, Thapar University, Patiala, India, 145(10), September, 2012.

[7] E. Ming-Yang Wu and Shu-Lung Kuo. A Study on the Use of a Statistical Analysis Model to Monitor Air Pollution Status in an Air Quality Total Quantity Control District. *Atmosphere*, 4:349–364, 2013.

[8] M. Chavent, H. Guegan, V. Kuentz, B. Patouille and J. Saracco. PCA and PMF based methodology for air pollution sources identification and apportionment. *Environmetrics*, 20:928–942, 2009.

[9] D. Voukantsis, K. Karatzas, J. Kukkonen, T. Räsänen, A. Karppinen and M. Kolehmainen. Inter-comparison of air quality data using principal component analysis, and forecasting of PM10 and PM2.5 concentrations using artificial neural networks, in Thessaloniki and Helsink. *Science of The Total Environment*, 409(7)1266–1276, 2011.

[10] J. G. Monroy, J. G.-Jimenez and J. L. Blanco. Overcoming the Slow Recovery of MOX Gas Sensors through a System Modeling Approach. *Sensors*, 2012.

[11] C. Wang, L. Yin, L. Zhang, D. Xiang and R. Gao. Metal Oxide Gas Sensors: Sensitivity and Influencing Factors. *Sensors*, 10(3):2088–2106, 2010.

[12] I. Morsi. Electronic Nose System and Artificial Intelligent Techniques for Gases Identification. *Arab Academy for Science and Technology*, Electronics and Communications Department Alexandria, Egypt. 2010

[13] W. Yan, Z. Tang, G. Wei and J. Yang. An Electronic Nose Recognition Algorithm Based on PCA-ICA Preprocessing and Fuzzy Neural Network. School of Electronic Science and Technology, Dalian University of Technology, Dalian, P.R. China, 2012.

Biographies

Souhir Bedoui was born in Tunisia, in 1987. She received the Electronic Engineer Diploma in 2011, a Master degree in electronic in 2012 then a PhD in electric in 2016 from National Engineer School of Sfax, Tunisia. The fields of microelectronics and micro-gas sensors are the research axes that she chose during her in-depth studies at the Master's and Doctorate.

Hekmet Charfeddine Samet was born in Sfax (Tunisia) in 1956. She obtained the Engineering Diploma and a PhD in physical sciences from National school of Engineering of Sfax (ENIS) and the accreditation to supervise research in electronic, respectively in 1983, 1997 and 2010. Currently, she is an associate Professor at ENIS School of Engineering and a member in the "LETI" Laboratory.

Abdennaceur Kachouri was born in Sfax, Tunisia, in 1954. He received the engineering diploma from National school of Engineering of Sfax in 1981, a Master degree in Measurement and Instrumentation from National school of Bordeaux (ENSERB) of France in 1981, a Doctorate in Measurement and Instrumentation from ENSERB, in 1983. He works on several cooperation with communication research groups in Tunisia and France. Currently, he is Permanent Professor at ENIS School of Engineering and member in the "LETI" Laboratory, ENIS Sfax.

H. Al-Libawy, A. Al-Ataby, W. Al-Nuaimy and M. A. Al-Taee

Multisensor Data Fusion in Fatigue Detection Using Wearable Devices

Abstract: Operator performance and safety that are both affected by the operator mental status (fatigue/alert) are basic requirements in work environments. The needs for practical and low-cost approaches for fatigue detection are therefore required by governmental, industrial and safety organizations. This paper proposes a new approach for operator fatigue detection that is based on biological data collection using accurate, low-cost and easy to use wearable devices. Three bio-data sensors for heart rate, wrist temperature and skin conductivity are adopted in this work for data collection and generation of fatigue-related metrics. Effective features of the collected bio-data are identified and labeled using the heart-rate variability metric that is measured by a wearable chest-strap heart monitor. The data collected from real subjects is used to train a dataset for fatigue analysis and classification using sub-classifiers based on artificial neural networks. Decision-level data fusion technique based on Bayesian combiner is then applied to enhance the accuracy and confidence of the obtained classification results. Performance of the developed alertness/fatigue detector is assessed experimentally and the obtained findings demonstrated acceptable performance in terms of modularity, accuracy, sensitivity and specificity when compared to individual classifiers.

Keywords: alertness; artificial neural networks; heart-rate variability; operator fatigue; Bayesian fusion; wearable sensors

1 Introduction

Fatigue is a mental state and usually combined with slower response times. The circadian rhythm and sleep deprivation are the drives of this natural state. The importance of research in mental fatigue problem is the fatal errors that come from operators and drivers that may lead to dangerous consequences [1]. For example, air transportation is one of the fields that is thoroughly covered for risk assessment; including fatigue and sleepiness as main risk factors [2]. Although some studies suggest that aviation transportation is the safest form of transportation, human errors remain the main cause of accidents [3].

Different approaches and methods have been reported in literature to study the operator fatigue. Most of these approaches were using biological laboratory data

H. Al-Libawy, A. Al-Ataby, W. Al-Nuaimy and M. A. Al-Taee: Department of Electrical Engineering and Electronics University of Liverpool, Liverpool, UK, , Emails: abbood@liverpool.ac.uk, aliataby@liverpool.ac.uk, wax@liverpool.ac.uk, altaeem@liverpool.ac.uk

De Gruyter Oldenbourg, ASSD – Advances in Systems, Signals and Devices, Volume 6, 2018, pp. 181–196.
https://doi.org/10.1515/9783110448375-012

collected by relatively expensive medical equipment. Numerous machine learning and classification algorithms such as Artificial Neural Networks (ANN), Support Vector Machine (SVM), K-nearest Neighbour (KNN) and others have also been proposed and used in a wide range of applications including fatigue detection [4–6], medical diagnosis [7–9], decision support and therapy of chronic diseases [10] and other applications. However, performance of these algorithms can vary from one application to another [11]. In [12], the fluctuation in heart rate was utilized for fatigue detection and monitoring. The heart rate volatility that does not represent the heart-rate variability (HRV) but relevant to it was obtained from historical data of the heart rate fluctuation. The relationship between frequency power ranges of HRV in the frequency domain components was reported in [13]. It was described that the ratio of low to high frequency components is inversely proportional with fatigue evolution (i. e. high ratio for low fatigue level and vice versa).

Biological explanation of the HRV behaviour is based on the fact that the activity of the autonomic nervous system which is addressed as a trusted source of information. It has two main components; sympathetic nervous system and parasympathetic nervous system [5]. The interaction between those two components is reflected in some biological signs; of these, the heart rate, core temperature and skin conductivity are the most important. The HRV is significantly affected by the activity of the autonomic nervous system components that are in turn vary with sleep/wake activity [14–17].

Recent technology advancements have led to design and development of numerous low-cost wearable devices [18–20] capable of accurately measuring and collecting various biological data, including the HRV. Utilization of these devices in fatigue-related studies has become affordable and quite feasible over the past five years. Benefiting from these technology advances in wearable biosensors, this paper builds on and extends the work reported by the authors in [21]. It proposes a new approach for operator fatigue detection that is based on biological data collection using accurate, low-cost and easy to use wearable devices. A decision-level data fusion technique based on Bayesian filter is suggested to enhance the accuracy and confidence of the obtained classification results. Performance of the developed fatigue detector can therefore be favourable compared to the state of the art findings.

The rest of this paper is organized as follows. Section 2 overviews common metrics of the HRV, focusing on the ratio of low/high frequencies of the heart rate changes that are of a particular interest in this study and gives theoretical background of Bayesian theorem. Section 3 presents the materials and method of this study with a particular focus on implementation of the proposed fatigue detection approach. The obtained results are presented and discussed in Section 4. Finally, the work is concluded in Section 5.

2 Background

2.1 Heart rate variability

Heart-rate variability (HRV) can be defined with the aid of Fig. 1 as follows. The duration between two heartbeats (also called normal-to-normal interval, NN) is typically measured from two adjacent QRS complexes that are captured from the ECG signal [22]. It should be mentioned here that the RR and NN terms are interchangeably used in literature and clinical practice [23]. Figure 1 shows an example of an ECG signal with some details about QRS complex and RR intervals. The variation in RR intervals that represents the HRV is characterized as a non-intrusive technique. It can be practically used to measure the sympathetic and parasympathetic modulation in humans [24].

HRV metrics that are calculated from the RR periods that reflect the variation between heartbeats intervals can be categorized into five domains: time, frequency,

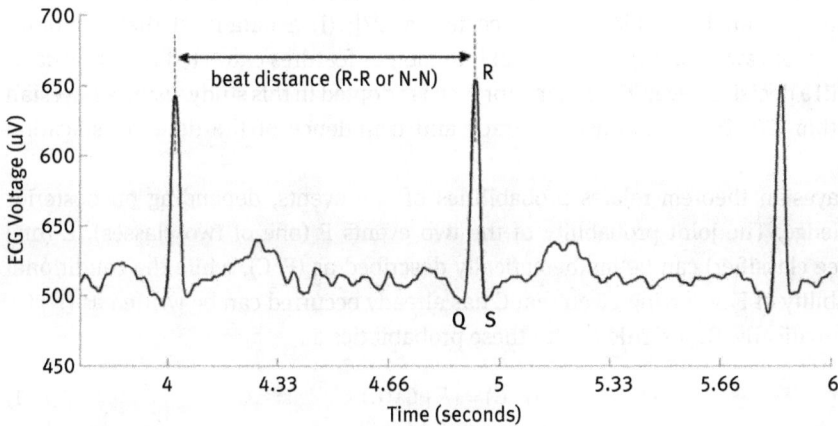

Fig. 1. Example of ECG signal with QRS complex.

Tab. 1. HRV Metrics.

Domain	Metric	Description
Time domain	SDNN	Standard Deviation of NN intervals
	RMSSD	Root Mean Square of Successive Differences
	NN50	Number of pairs of successive NNs that differ by > 50 ms
Frequency domain	VLF	Very Low Frequency power from 0.0033–0.04 Hz
	LF	Low Frequency power from 0.04–0.15 Hz
	HF	High Frequency power from 0.15–0.4 Hz
	LF/HF	Ratio of low to high frequency power

complexity, fractal and nonlinear [25]. A measurement period of at least 5 minutes is considered to calculate the HRV from ECG, as suggested in [26]. Table 1 summarizes the metrics of common use in the time and frequency domains. Of these, the LF/HF is considered of a particular importance in this study.

2.2 Bayesian data fusion

Several definitions for data fusion have been reported in literature. However, the mostly agreed upon definition is [27], *"Information fusion is the synergistic integration of information from different sources about the behaviour of a particular system, to support decisions and actions relating to the system."*

In this study, a dataset is created through collecting data from volunteer participants through several bio-data sensors (i. e. heart rate, wrist temperature and skin conductivity). As the raw data collected from these sensors cannot be merged directly, a data fusion technique is proposed and implemented to improve overall performance of the proposed fatigue detector. Generally, data/information fusion approaches can be divided into three levels [27]; (i) a data-level that combines multisensor raw data, (ii) feature-level that merges features extracted from raw data, and (iii) a decision-level. The latter approach is adopted in this study, using a Bayesian algorithm [28, 29] to improve accuracy and confidence of the data classification stage.

Bayesian theorem relates probabilities of two events, depending on posterior knowledge. The joint probability of the two events F (one of two classes), C (one of three classifier) can be mathematically described as (F, C), while the conditional probability of F occurring given that C has already occurred can be written as (F | C). Mathematically, Bayes' rule relates these probabilities as

$$(F, C) = (F \mid C)(C) \tag{1}$$

$$(F \mid C) = \frac{(C \mid F)(F)}{(C)} \tag{2}$$

If there are several events F_i, then event F can be written as normalization of mutually exclusive events as:

$$(F \mid C) = \frac{(C \mid F)(F)}{\sum_i (C \mid F_i)(F_i)} \tag{3}$$

The term $(C_i \mid F_i)$ can be calculated from individual ANN classifier outputs as prior probabilities [30] and the classifier confidence can be approximated as a posterior probability [31]. Taking into account these assumptions, Bayesian fusion can be implemented to enhance the overall classifier performance post the fusion process.

3 Materials and method

A total of 9 male volunteers, aged 16–50 years with body mass indexes of 21–35 were participated in this study. Each participant is provided with two wearable devices; a fitness tracker watch (shown in Fig. 2a) and a heart-rate sensor strap (Polar H7, shown in Fig. 2b). The fitness tracker watch is capable of saving bio-data in its internal memory is adopted in this study. It is used to collect several bio-data, including heart rate, body temperature, and skin conductance. The participant's data was collected at a rate of 1 reading per minute.

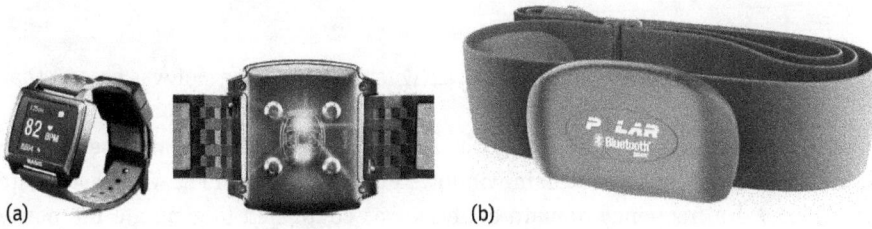

Fig. 2. Examples of a fitness wearable devices. (a): Fitness tracker watches [32]. (b): Fitness heart-rate sensor [33].

The HRV is obtained from the RR period is measured by heart-rate sensor strap that is equipped with a Bluetooth communication facility. The measured data can therefore be wirelessly transmitted in real time to a wide variety of handheld devices (e. g. smartphones and tablets). The data is collected around 16 hours daily using the fitness heart-rate sensor strap in companion with a smartphone application to collect and record data from sensors and then upload it to a remote website, Fluxtream [34].

The method of this study comprises three distinct stages; (i) pre-processing, (ii) feature extraction and labelling and (iii) fatigue detection. These stages are shown in the block diagram of Fig. 3 and are described as follows.

3.1 Pre-processing and HRV calculation

3.1.1 Pre-processing

Missing and out of range data are common problems in data collection; it can be caused by a sensor or a human failure. For example, a participant inability of wearing the watch or the chest strap during battery charging or shower time. To mitigate the impact of this problem, some practical arrangements are considered; (i) dealing with slots of less than 30 samples through interpolating the missing data and (ii) using

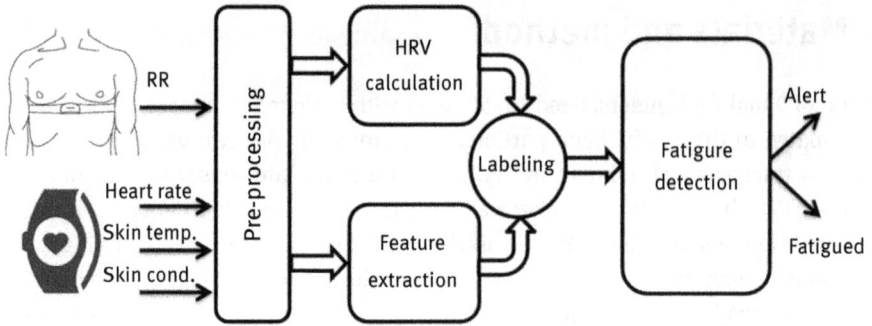

Fig. 3. Block diagram for the proposed system.

unequally space frequency domain analysis when data-missing slot was greater than 30 samples.

The collected data from tracking watch was pre-processed and analysed to generate statistical metrics like 30-min windowed mean (shown in Fig. 4) and standard deviation also a frequency domain analysis was conducted to generate the power spectral density and calculate the three bands of power frequency. Finally, a set of several features were selected as an input-data vector. The selected set of features are discussed later in Section 3.2.

3.1.2 HRV measurement

An example of 10-minute RR intervals for one of the participants involved in this study is shown in Fig. 5. The collected RR data from chest strap is analysed and used to generate HRV. An average record duration of 10 mins is considered adequate to obtain reasonably accurate and reliable HRV metrics. Theses metrics are calculated with the aid of an existing HRVAS application reported in [35, 36].

Power Spectral Density (PSD) is usually calculated using many methods and analysis approaches. In this work, three methods are adopted [35, 37]; Welch, Burg and Lomb-Scargle. Figure 6 shows two examples of PSD in which we can notice that the ratio of LF/HF increases in the midday while decreases at late night. This fatigue-correlated change in LF/HF metric is used in this work and labelled as two classification states as alert and fatigued participant.

As expected, LF/HF metric shows a trend of a clear correlation with growth of fatigue at night and this metric is chosen to represent the output data. Figure 7 shows an example of one participant calculated using the above-mentioned methods. The calculated LF/HF metric around the wakening hours which demonstrates the growth of this metric from early morning, reaching its maximum at afternoon, and then

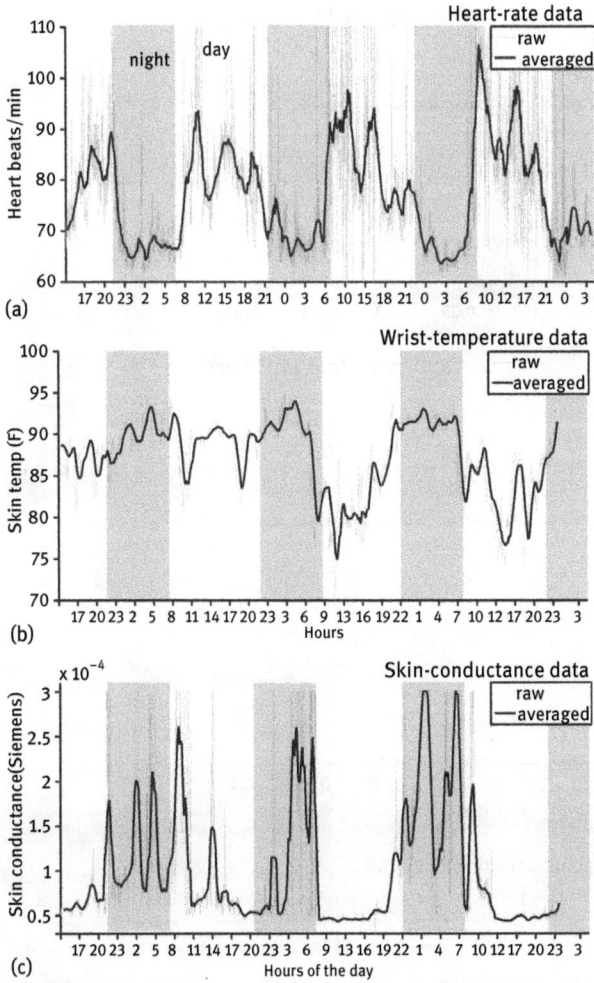

Fig. 4. Examples of the collected bio-signals.

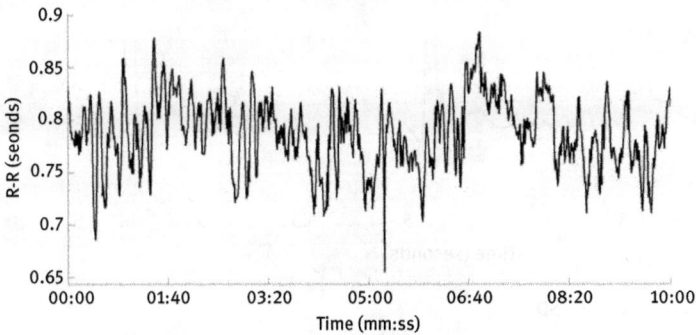

Fig. 5. RR intervals extracted from heart-rate monitor.

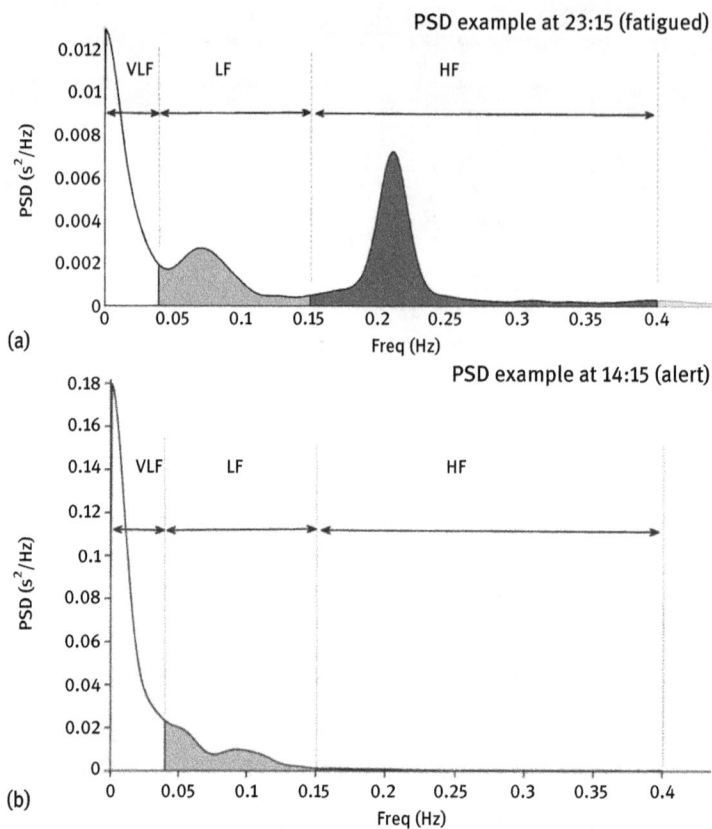

(a)

(b)

Fig. 6. Example record of power spectral density for HRV.

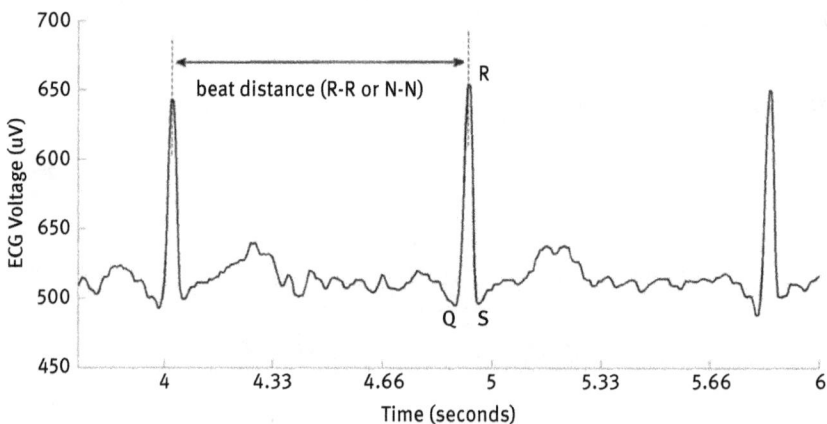

Fig. 7. Example of HRV metric, LF/HF PSD.

decaying at the end of day. Despite that the three PSD calculations methods show the alertness pattern, Welch and Burg methods tend to coincide all the times.

3.2 Feature extraction and data labelling

The collected data from the fitness watch (hear rate, wrist temperature and skin conductance) are passed to feature extraction stage after pre-processing stage. After feature extraction, the following 6 out of 14 features are chosen to be the most effective:
- Heart rate 30 sample windowed mean
- Heart rate standard deviation
- Wrist temperature 30 sample windowed mean
- Wrist temperature standard deviation
- Skin conductance 30-sample windowed mean
- Skin conductance standard deviation

Based on LF/HF measures that are calculated from RR data, smoothing and interpolation fitting is implemented over data to produce the output vector of the classifier. Figure 8 shows an example of a polynomial fitted HRV metric (LF/HF) for one participant calculated in three methods (Welch, Burg and Lomb-Scargle). The three curves showing in this figure represent the three method of PSD calculation (Welch, Burg and Lomb-Scargle).

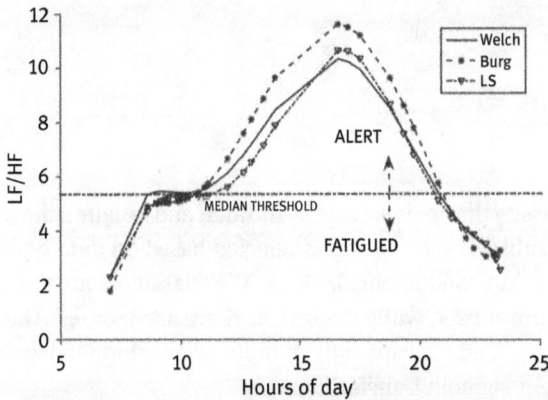

Fig. 8. Example of a polynomial fitted HRV metric, the LF/HF.

Among the PSD methods mentioned above,the Welch method was selected in this study. This was mainly because it is producing the highest correlation coefficient between the HRV fitting polynomial and pre-processed data. Dynamic threshold, which is depending on individual differences, is chosen to label the output vector into alert and fatigued states. The median metric of HRV data over the day is chosen as threshold level. This threshold is used to label the bio-data set with two labels, alert label for the part above the threshold line and fatigued for the part below the threshold line. The bio-data set are combined with its labels to create training set and use it with supervised machine learning algorithms to build fatigue classifier.

3.3 Fatigue detection

This stage proceeds in two steps; classification and fusion. The labelled features are fed to detection stage which in turn generates decision on the operator fatigue status (i. e. alert or fatigued). Figure 9 shows a block diagram for the proposed tow-steps detector, starting from three sub-classifiers and ending with Bayesian fusion stage.

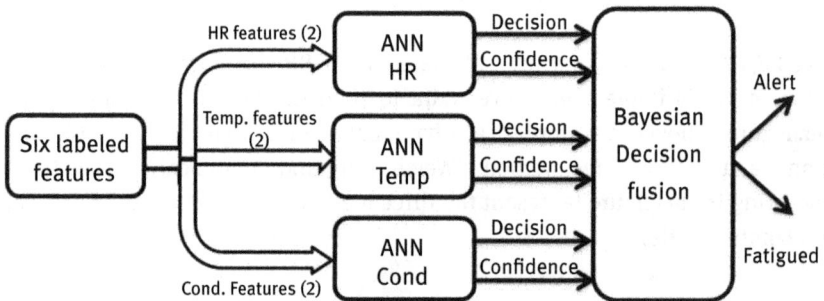

Fig. 9. Two-stage fatigue detector (classifier and fuser).

3.3.1 Classification

The collected dataset is used to classify the operator status into alert and fatigue states. The dataset is divided into three subsets; each subset is selected based on data type (heart rate, wrist temperature and skin conductance). Three ANN classifiers are then trained by 70 % of individual feature subset, while the rest 30 % are used for test. The structure of the ANNs is based on the feed-forward with an input layer, a hidden layer and two output units with a tangent-sigmoid transfer function.

3.3.2 Bayesian Fusion

At this stage, the final decision are generate using the decision and confidence values received from the ANNs. As shown in Fig. 9, the output of each classifier is approximated as posteriors probability. Bayesian fusion algorithm then combine the output of the sub-classifiers by applying maximum a posteriori probability (MAP) rule [38, 39]. The detection results post fusion stage are shown in Table 2. As illustrated, the overall classification accuracy, sensitivity and specificity are improved when compared to the results obtained from the sub-classifiers.

Tab. 2. Summary of system performance.

Output states	ANN classifiers			
	Heart-rate features	Wrist- temperature features	Skin- conductance features	Bayesian Fuser
Accuracy	71.80%	70.6%	62.20%	75.96%
Sensitivity	62.00%	68.50%	79.20%	67.76%
Specificity	80.60%	80.50%	59.80%	82.71%

4 Results and discussion

Numerous experiments have been carried out over a period of 9 weeks; (each experiment takes about a week to complete). The participants were instructed to collect and synchronize data with a remote server on daily bases. A user-friendly data collection and management application was deployed and used on the handheld device for this purpose.

Different configurations were also considered to identify the ANN structure with the best performance. These configurations involved changing the number of hidden layers and associated nodes as well as optimizing the training algorithm and the decision transfer function. Levenberg-Marquardt back-propagation algorithm was eventually selected for the ANNs' training. Figure 10 shows an example of the sub-classifiers performance in terms of two metrics; (i) confusion matrix and (ii) receiver operating characteristics.

Several trials of randomly selected records from data sets were conducted to calculate the accuracy of classification for all classifiers. Table 2 clearly shows the superiority of the Bayesian fuser results over the individual sub-classifiers in terms of accuracy and specificity. This trend of findings is also expected to be valid for most classifiers when they fed with larger set of effective features. The sub-classifiers demonstrated close results in terms of accuracy. The heart rate sub-classifier demonstrated the highest accuracy (71.8%), while the skin conductance sub-classifiers demonstrated the least accuracy (62.2%).

Confusion matrix

	Alert	**Fatigued**	
Alert	1062 42.1%	255 10.1%	80.6%
Fatigued	457 18.1%	747 29.6%	62.0%
	69.9%	74.6%	71.8%

Output Class

Alert   Fatigued
(a)   **Target Class**

Receiver Operating Characteristic

ROC

True Positive Rate vs False Positive Rate — Alert, Fatigue

(b)

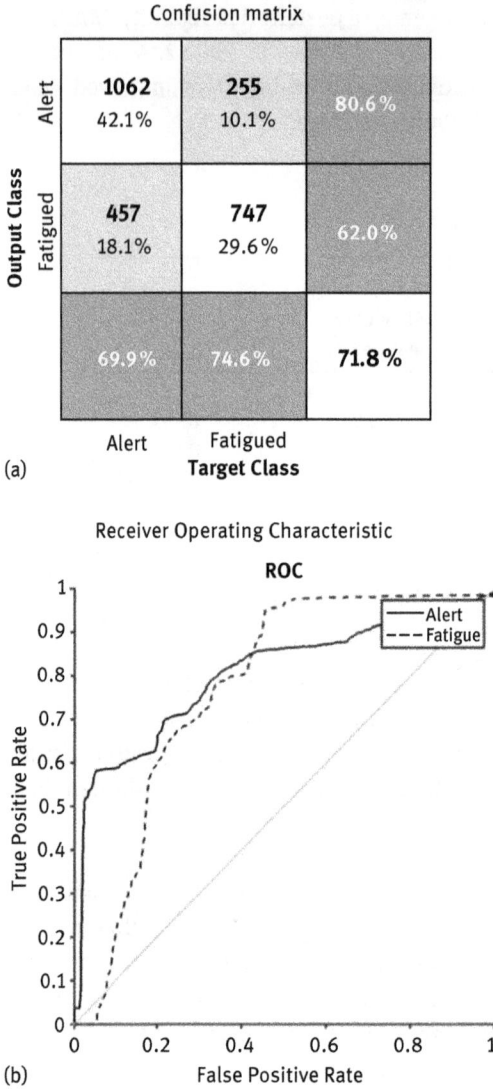

Fig. 10. Example of sub classifiers performance (heart-rate classifier).

5 Conclusions

A new multisensor fatigue detection system has been proposed and implemented successfully. The developed prototype was found to be promising in terms of usage low-cost wearable devices to detect fatigue status of operators in real-life environments with an acceptable level of accuracy. The classifiers and the fuser results demonstrated performance differences relevant to different sizes of dataset or different approaches. The fuser results showed the highest performance of accuracy when comparedto those obtained from the sub-classifiers. Moreover, the fusion algorithm is efficiently applied in modular and distributed system which can base on multi-subsystem with less computing power. The operator individuality is enhanced in labeling process by choosing different levels of threshold based on HRV median divider.

This study is still open for further improvements and verifications that may include but not limited to: (i) conducting a wider study which includes males and females, and different age groups, (ii) extending this study to consider more fusion levels (i. e. data and feature levels), and (iii) Considering more fatigue-related metrics such as behavioral and visual metrics. These improvements and others are currently part of the on-going work of the authors.

Acknowledgment: This research was supported by a PhD fellowship from Babylon University- Ministry of Higher Education and Scientific Research, Iraq; grant no. MOHESR-IQ-2013-1257.

Bibliography

[1] H. Abbood, W. Al-Nuaimy, A. Al-Ataby, S. A. Salem and H. S. AlZubi. Prediction of driver fatigue: Approaches and open challenges. 14th *UK Workshop on Computational Intelligence* (UKCI), Bradford, England, September, 8–10, 2014.

[2] M. Ingre, W. Van Leeuwen, T. Klemets, C. Ullvetter, S. Hough, G. Kecklund, D. Karlsson and T. Åkerstedt. Validating and extending the three process model of alertness in airline operations. *PLoS ONE*, 9:e108679, 10, 2014. http://dx.doi.org/10.1371%2Fjournal.pone.0108679

[3] R. Hall. "Transportation science. *Handbook of Transportation Science*, ser. International Series in Operations Research and Management Science, R. Hall, Ed., Springer US, 2003, 56:1–5. http://dx.doi.org/10.1007/0-306-48058-1_1

[4] H. Al-Libawy, A. Al-Ataby, W. Al-Nuaimy and M. A. Al-Taee. Fatigue detection method based on smartphone text entry performance metrics. 9th *Int. Conf. on Developments in eSystems Engineering* (DeSE 2016), Liverpool & Leeds, England, August 31 – September 2, 2016.

[5] N. Munla, M. Khalil, A. Shahin and A. Mourad. Driver stress level detection using HRV analysis. *Int. Conf. on Advances in Biomedical Engineering* (ICABME), :61–64, September, 2015.

[6] H. Al-Libawy, A. Al-Ataby, W. Al-Nuaimy and M. A. Al-Taee. HRV-based operator fatigue analysis and classification using wearable sensors. 14th *Int. Multi-Conf. on Systems, Signals & Devices* (SSD), Marrakech, Morocco, March 28–31, 2017.

[7] A. Y. Al-Hyari, A. M. Al-Taee and M. A. Al-Taee. Diagnosis and classification of chronic renal failure utilising intelligent data mining classifiers. *Int. Journal of Information Technology and Web Engineering* (IJITWE), 9(4):1–12, 2014.

[8] M. Al-Taee, A. Z. Zayed, S. N. Abood, M. A. Al-Ani, A. M. Al-Taee and H. A. Hassani. Mobile-based interpreter of arterial blood gases using knowledge-based expert system. *Int. Journal of Pervasive Computing and Communications*, 9(3):270–288, 2013.

[9] A. Al-Taee, A. Al-Taee, Z. Muhsin, M. Al-Taee and W. Al-Nuaimy. Towards developing online compliance index for self-monitoring of blood glucose in diabetes management. 9th *Int. Conf.*

on Developments in eSystems Engineering (DeSE), Liverpool & Leeds, England, August 31 – September 2, 2016.

[10] A. M. Al-Taee, M. A. Al-Taee, W. Al-Nuaimy, Z. J. Muhsin and H. AlZu'bi. Smart bolus estimation taking into account the amount of insulin on board. *IEEE Int. Conf. on Computer and Information Technology; Ubiquitous Computing and Communications; Dependable, Autonomic and Secure Computing; Pervasive Intelligence and Computing* (CIT/IUCC/DASC/PICOM), :1051–1056, 2015.

[11] A. Y. Al-Hyari, A. M. Al-Taee and M. A. Al-Taee. Clinical decision support system for diagnosis and management of chronic renal failure. *IEEE Jordan Conf. on, Applied Electrical Engineering and Computing Technologies* (AEECT), :1–6, 2013.

[12] W. He. Application heart rate variability to driver fatigue detection of dangerous chemicals vehicles. 5th *Int. Conf. on Intelligent Systems Design and Engineering Applications* (ISDEA), :218–221, 2014.

[13] G. D. Furman, A. Baharav, C. Cahan and S. Akselrod. Early detection of falling asleep at the wheel: A heart rate variability approach. *Computers in Cardiology*, :1109–1112, 2008.

[14] M. Bonnet and D. Arand. Heart rate variability: sleep stage, time of night, and arousal influences. *Electroencephalography and Clinical Neurophysiology*, 102(5):390–396, 1997.

[15] R. D. Ogilvie. The process of falling asleep. *Sleep Medicine Reviews*, 5(3):247–270, 2001.

[16] Z. Shinar, S. Akselrod, Y. Dagan and A. Baharav. Autonomic changes during wake–sleep transition: A heart rate variability based approach. *Autonomic Neuroscience*, 130(1):17–27, 2006.

[17] W. Karlen, C. Mattiussi and D. Floreano. Sleep and wake classification with ecg and respiratory effort signals. *IEEE Trans. on Biomedical Circuits and Systems*, 3(2):71–78, 2009.

[18] M. Mateu-Mateus, F. Guede-Fernández and M. A. García-González. Rr time series comparison obtained by h7 polar sensors or by photoplethysmography using smartphones: breathing and devices influences. 6th *European Conf. of the Int. Federation for Medical and Biological Engineering*, Springer, :264–267, 2015.

[19] H. Al-Libawy, W. Al-Nuaimy, A. Al-Ataby and M. A. Al-Taee. Estimation of driver alertness using low-cost wearable devices. *Conf. on Applied Electrical Engineering and Computing Technologies* (AEECT), Jordan, :1–5, 2015.

[20] F. Guede-Fernandez, V. Ferrer-Mileo, J. Ramos-Castro, M. Fernandez-Chimeno and M. Garcia-Gonzalez. Real time heart rate variability assessment from android smartphone camera photoplethysmography: Postural and device influences. 37th *IEEE Annual Int. Conf. on Engineering in Medicine and Biology Society* (EMBC), :7332–7335, 2015.

[21] H. Al-Libawy, A. Al-Ataby, W. Al-Nuaimy and M. A. Al-Taee. HRV-based operator fatigue analysis and classification using wearable sensors. 13th *Int. Multi-Conf. on Systems, Signals & Devices*, Leipzig, Germany, March 21–24, :268–273, 2016.

[22] M. Aqel and M. Al-Taee. Mobile-based interpreter of arterial blood gases using knowledge-based expert system. *Journal of Advances in Modelling (B) - Signal Processing and Pattern Recognition*, 45(4):61–71, 2002.

[23] A. Bonjyotsna and S. Roy. Correlation of drowsiness with electrocardiogram: A review. *Int. Journal of Advanced Research in Electrical, Electronics and Instrumentation Engineering*, 3(5), 2014.

[24] A. Noda, M. Miyaji, Y. Wakuda, Y. Hara, T. Yasuma, F andFukuda, K. Iwamoto and N. Ozaki. Simultaneous measurement of heart rate variability and blinking duration to predict sleep onset and drowsiness in drivers. *Journal of Sleep Disorders & Therapy*, 4(5), 2015. http://www.omicsgroup.org/journals/simultaneous-measurement-of-heart-rate-variability-and-blinking-duration-to-predict-sleep-onset-and-drowsiness-in-drivers-2167-0277-1000213.php?aid=59828

[25] S. Ahmad, M. Bolic, H. Dajani, V. Groza, I. Batkin and S. Rajan. Measurement of heart rate variability using an oscillometric blood pressure monitor. *IEEE Trans. on Instrumentation and Measurement*, 59(10):2575–2590, 2010.

[26] J. McNames and M. Aboy. Reliability and accuracy of heart rate variability metrics versus ecg segment duration. *Medical and Biological Engineering and Computing*, 44(9):747–756, 2006.

[27] International society of information fusion (isif). http://isif.org/, accessed: 2017-04-20.

[28] F. Roli. Multiple classifier systems. *Encyclopedia of Biometrics*, :1142–1147, 2015.

[29] B. Bigdeli, F. Samadzadegan and P. Reinartz. A decision fusion method based on multiple support vector machine system for fusion of hyperspectral and lidar data. *Int. Journal of Image and Data Fusion*, 5(3):196–209, 2014.

[30] D. W. Ruck, S. K. Rogers, M. Kabrisky, M. E. Oxley and B. W. Suter. The multilayer perceptron as an approximation to a bayes optimal discriminant function. *IEEE Trans. on Neural Networks*, 1(4):296–298, 1990.

[31] G. Papadopoulos, P. J. Edwards and A. F. Murray. Confidence estimation methods for neural networks: A practical comparison. *IEEE Trans. on Neural Networks*, 12(6):1278–1287, 2001.

[32] PEAK - Fitness and Sleep Tracker Watch, http://www.mybasis.com, [Online; accessed 21-Oct-2015].

[33] Polar H7 heart rate sensor strap, http://www.polar.com/uk-en/products/accessories/H7_heart_rate_sensor, [Online; accessed 21-Apr-2017].

[34] Fluxtream: open source non-profit personal data visualization framework, https://www.fluxtream.org/, [Online; accessed 21-Apr-2017].

[35] J. Ramshur. *Design, evaluation and application of heart rate variability software (HRVAS)*. Master's thesis, The University of Memphis, Memphis, TN, 2010.

[36] X. Xu. *Analysis on mental stress/workload using heart rate variability and galvanic skin response during design process*. Master's thesis, Concordia University, 2014.

[37] P. Stoica and R. L. Moses. *Spectral analysis of signals*. Pearson Prentice Hall Upper Saddle River, NJ, 452, 2005.

[38] B. Le Saux and H. Bunke. Combining svm and graph matching in a bayesian multiple classifier system for image content recognition. *Joint IAPR Int. Workshops on Statistical Techniques in Pattern Recognition (SPR) and Structural and Syntactic Pattern Recognition (SSPR)*. Springer, :696–704, 2006.

[39] C. De Stefano, C. D'Elia, A. Marcelli and A. S. di Freca. Using bayesian network for combining classifiers. 14th *Int. Conf. on Image Analysis and Processing*, (ICIAP), :73–80, 2007.

Biographies

Hilal Al-Libawy received BSc degree in Electrical Engineering from Baghdad University, Baghdad, Iraq, in 1991, MSc degree in electronic engineering in 1995. He is a teaching staff in Babylon University, Babylon, Iraq since 2004 till now. Al-Libawy a PhD student in behavourial analysis and operator fatigue studies since 2013 in University of Liverpool, Liverpool, UK. His main areas of research interest are operator fatigue detection, machine learning, and biological and cognitive modelling including ACT-R architecture.

Ali Al-Ataby received BSc degree in Electronic and Telecommunications Engineering from Baghdad, Iraq, in 1997, and subsequently MSc degree in electronic circuits and systems engineering in 1999. In 2012, he received a Ph.D. degree in electrical engineering from the University of Liverpool, UK. He joined the Department of Electrical Engineering and Electronics at the University of Liverpool as a lecturer in signal processing in 2011. He worked in academia from 1997 to 2002 and in industry from 2002–2009. Current research interest is centered on devising automatic interpretation algorithms for non-destructive testing data, with a particular interest in visual, ultrasonic and radar data. Further interests are in biomedical signal/data processing (e. g. EEG, ECG and DNA data), and in driver fatigue detection/management. Other research interests include machine learning, hardware signal processing and microcontroller/microprocessor based embedded systems.

Waleed Al-Nuaimy received his BSc in Electronic and Telecommunications Engineering from Baghdad, Iraq, in 1995 and then completed his PhD at the University of Liverpool, UK, in 1999. Since then he has worked at the Department of Electrical Engineering and Electronics in the signal Processing research group. His main research areas are automated analysis of data, autonomous systems and human computer interaction. His other related interests involve the automated analysis of nondestructive testing and biomedical data, and machine learning for behavioral analysis.

Majid A. Al-Taee received his BSc in Control and Systems Engineering from the University of Technology, Baghdad, Iraq in 1984 and PhD in Electrical Engineering and Electronics from the University of Liverpool, UK, in 1990. Prior to joining Kingston University London, UK in 2012, and the University of Liverpool in 2014, he was a full Professor and Vice Dean of the Faculty of Engineering and Technology at the University of Jordan. He received numerous research funding awards, scientific prizes, and published more than 120 papers in peer-reviewed international journals/conferences. Prof Al-Taee is a senior member of the IEEE, and Chair/Vice Chair of the IEEE Computer/Computational Intelligence Chapter-Jordan Section (2004–2012). Also, he is a member of the Marie Curie Alumni Association and member of the editorial board of the International Journal of Digital Signals and Smart Systems. His research focuses on distributed and cloud computing, electronic/mobile health systems, machine intelligence, and electrical systems control and stability analyses.

M. Dorwarth, S. Kehrberg, R. Maul, R. Eid, F. Lang, B. Schmidt
and J. Mehner

Application of a Trajectory Piecewise Linearization Approach on a Nonlinear MEMS Gyroscope

Abstract: Despite continuously increasing computing speed of modern calculation clusters, the transient simulation of system models with nonlinear effects is still challenging. This is especially valid in the context of highly complex structures such as MEMS gyroscopes. Model order reduction (MOR) methods such as the modal superposition have been successful in the description of linear systems, but cannot be used to simulate nonlinearities. Here, a trajectory piecewise linearization (TPWL) is presented and its application on nonlinear mechanics is demonstrated. Using the TPWL on dynamic MEMS high Q systems, in particular MEMS gyroscopes, is innovative and relevant for development processes. In the following, the computation time of a transient one mass oscillator system could be reduced by a factor of 1000 with the TWPL, while the physics of the system are preserved with an agreement of more than 95 % to a reference FEM simulation. Furthermore, a test application of the TPWL on an especially nonlinear designed MEMS gyroscope is outlined. This discussion includes convergence studies on the nonlinear TPWL eigenfrequencies and shape analyses of the corresponding modes. Moreover, different approaches of TPWL on this system are compared for a nonlinear frequency shift over deflection – one of them leading to less than 11 % discrepancy to measurements and less than 3 % discrepancy to FEM reference simulations.

Keywords: MEMS Gyroscope, TPWL, MOR, high Q system, nonlinearities

1 Introduction

The MEMS gyroscope is one of the most challenging micromechanical devices in the context of simulation methodology. Especially the highly complex MEMS core structure leads to movement patterns which can be difficult to predict. On the other hand, an accurate description is needed to gain a more fundamental understanding of the system and its nonlinear behavior, leading to an optimization of the development process for future, more advanced gyroscopes. An insight into the complexity of

M. Dorwarth, S. Kehrberg, R. Maul, R. Eid, F. Lang and B. Schmidt: M. Dorwarth, S. Kehrberg,
R. Maul, R. Eid, F. Lang and B. Schmidt: Robert Bosch GmbH, Stuttgart, Germany, e-mails:
Markus.dorwarth@de.bosch.com.
J. Mehner: Department of Microsystems and Precision Engineering, Chemnitz University of
Technology, Chemnitz, Germany, e-mails: Jan.mehner@etit.tu-chemnitz.de.

De Gruyter Oldenbourg, ASSD – Advances in Systems, Signals and Devices, Volume 6, 2018, pp. 197–216.
https://doi.org/10.1515/9783110448375-013

a MEMS gyroscope functional principle and structure has been given in several publications (e. g. [1, 2]) and literature (e. g. [3]).

Transient simulation methods are especially relevant in the development process of MEMS gyroscopes. The finite element model (FEM) tools as e. g. ANSYS® or Abaqus® are well established for static simulations of MEMS gyroscopes for many years and are very precise as well (see e. g. [4]), but they are too demanding for applications on dynamical complex systems (see e. g. [5]). Consequently, other methodologies are favored. Model order reductions (MOR), in which the system's degrees of freedom (DOF) are reduced from more than 10^6 to the order of 10 are a common way to simulate time dependent models. Several promising results have been achieved with these approaches [6, 7].

Full system simulations, including electromagnetic forces and a regulating ASIC, are even more challenging to compute with FEM tools than the mechanical structure alone. In particular, adding the ASIC to the model is difficult, because the simulation of its output can hardly be achieved by fragmenting it into finite elements. Instead, electronic design automation (EDA) software is used here (see e. g. [8]). A MOR model on the other hand can be constructed in Matlab Simulink [5]. This tool allows, with some limitations regarding Kirchhoff's circuit laws, the emulation of an ASIC based on signal flow simulations. Therefore, full system computations are easier to built here [5, 9]. A common MOR method in the field of mechanical applications and MEMS is the modal superposition [6, 7]. Unfortunately, this strategy just covers linear mechanical effects and neglects geometric nonlinear influences caused by stress stiffening of the sensor structure, which leads to unforeseen excitations and phase shifts of the gyroscope's mechanical eigenmodes. A well known approximation for the description of these nonlinearities is the Duffing equation, which yields to shark fin shaped resonance peaks [10, 11]. Also electrostatical nonlinearities can be respected in a signal flow simulation. The latter one is widely outlined in e. g. [12]. A very detailed discussion on the relevance of both, material and geometrical nonlinearities in silicon structures, is given in [13].

In other fields, nonlinear MORs based on a trajectory piecewise linearization (TPWL) are more established (e. g. electromagnetism [14], circuit simulation [15], flow modeling [16]). The first reference to this method can be found in [17], more detailed error estimations of this approach are given in [18]. In contrast to the mechanical equations of motion, the differentials in these systems are of first order and consequently, the realization is different. Furthermore, the used FEM software (still needed to build up a TPWL model) is often non-commercial or the implementation is not outlined (as e. g. in [17, 18]). While the nonlinear analysis of MEMS oscillators (i. e. calculation of frequency shifts over amplitude) is quite investigated (e. g. [19]), the transient system simulations are rare to be found in literature. There have only been a few approaches to use TPWL on mechanical MEMS systems, but these were no high Q models such as gyroscopes. Also, the underlying equations weren't the full system differential equations – instead, e. g. Euler's beam equation was used as in [17, 18].

Other methods than the TPWL have been applied on mechanical high Q MEMS models, e. g. as shown in [20, 21]. The given examples are a simple expansion of the modal superposition and therefore, lead to some simplifications. A concrete disadvantage is that the eigenvectors of this approach are equal to those of the linear system – this is an important qualitative difference to the TPWL which is not based on linear eigenmodes. Referring to the influence of high Q factors in MEMS: This also has been discussed before in [22]. It must be clear that the dynamical behavior of such systems is of special interest in the MEMS development and needs to be analyzed in great detail.

In the following, a nonlinear MOR for mechanical systems based on a TPWL methodology is introduced. The TPWL workflow, the implementation of the MOR and an application on a simple MEMS structure are discussed. The description of a MEMS high Q system with TPWL and a commercial FEM software is shown. Afterwards, a test application on an especially nonlinear designed MEMS gyroscope is outlined. Here, qualitative and quantitative behavior of the MOR simulation concerning input parameters and settings are discussed.

2 Methods

2.1 Equation of motion

In the finite element theory, all mechanical systems are described with a certain number of point masses, coupled by mechanical springs. The DOF per point mass depend on the chosen element type – apart from displacements rotational DOF, temperature, pressure and others are common.

For a description of the FEM system a p-dimensional differential equation has to be solved, where p is the total number of DOF:

$$M\ddot{x} + D\dot{x} + K(x)x = F(t). \tag{1}$$

Here, x is the nodal displacement vector, M the mass matrix, D the damping matrix, $K(x)$ the (nonlinear) stiffness matrix, $K(x)x$ the restoring or internal forces and $F(t)$ the applied load. In a linear simplification the stiffness matrix is constant and a modal superposition gives a good description of the system. As outlined above, this approach is a simplification and cannot describe nonlinear effects.

In nonlinear analyses, FEM tools such as ANSYS® or Abaqus® determine a first order solution of every time step n with the Newton-Raphson procedure [23]. Here, the residual of equation (1) is minimized iteratively. In this process, FEM tools do not derive the explicit function of the stiffness matrix. Instead, the internal load is re-calculated in each load step from the element information, i. e. a linearization of the system for all displacements. In transient simulations of systems with many nodes this is time consuming but necessary for precision.

As a consequence, the explicit form of the stiffness matrix cannot be exported from FEM analyses and used in MOR concepts. A different approach has to be employed: the TPWL.

2.2 Trajectory piecewise linearization

The TPWL follows a discrete linearization concept which is shown schematically in Fig. 1. The approach is based on the idea that the linearized information of each point in the p-dimensional space is a good approximation of the system close to this state. Therefore, equation (1) can be written as:

$$M\ddot{x} + D\dot{x} + f_{\text{lin}}(x_n) + K_{\text{lin}}(x_n)(x - x_n) = F(t) \qquad (2)$$

for all points x close to x_n. The matrix K_{lin} is the tangential matrix and the sum $f_{\text{lin}}(x_n) + K_{\text{lin}}(x_n)(x - x_n)$ is the first order taylor series of $K(x)x$ [24].

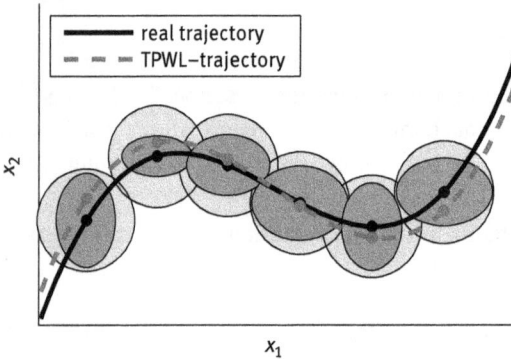

Fig. 1. The black curve is the trajectory of a real system used to create the training data. The dots on the black curve show the snap shot points and the circles illustrate the related validity region of the linearization at these states. The ellipses in purple demonstrate the smaller validity region of the reduced systems. The red curve gives the trajectory of the MOR approach.

The validity region of equation (2) can be extended by using an interpolation concept for the sum $f_{\text{lin}}(x_n) + K_{\text{lin}}(x_n)(x - x_n)$ [17]. A general form of this approach is given by

$$M\ddot{x} + D\dot{x} + \sum_{n=1}^{m} \{w_n(x)\,[f_{\text{lin}}(x_n) + K_{\text{lin}}(x_n)(x - x_n)]\} = F(t). \tag{3}$$

The $w_n(x)$ are the interpolation prefactors. The accuracy of this description depends on the number of interpolation points m, their position in p-dimensional space and the chosen interpolation concept.

Even without the MOR, interpolating between the linearization points increases the simulation speed considerably compared to an FEM approach.

Equation 3 can be reduced with a proper orthogonal decomposition (POD) as suggested in [16, 25] or a similar model order reduction method – an explanation of the POD method itself can be found in [26]. Alternative approaches of TPWL combined with the Krylow-subspace method and with the TBR have been shown as well [17, 27]. The choice of the MOR depends on the system, but in all cases the region of validity for a dynamical computation is reduced because a certain amount of information is lost. This effect is indicated in Fig. 1 by purple ellipses.

From equation (3) it can be seen that the $f_{\text{lin}}(x_n)$ and $K_{\text{lin}}(x_n)$ have to be determined in a previous analysis. This is done in an FEM tool by simulating a representative nonlinear trajectory of the system. The necessary information is exported and stored at the relevant time steps. This simulation is called *training run* or simply *training*. The steps at which data is exported are referred to as *snap shots* or *training points*.

Although the training run requires calculation time, a high benefit can be achieved if the derived MOR system is used several times afterwards e. g. as part of a system simulation.

2.3 Interpolation and time integration concept

The choice and the implementation of an adequate interpolation scheme is crucial for a high accuracy TPWL system model. Very common is the usage of a simple weight function (e. g. [14]). Here, all the reduced K_{lin} matrices and f_{lin} vectors are stored together with the order reduced displacement information q of the corresponding snap shot point where the latter may be in full or reduced coordinates. The p-dimensional distance between the training points and the current displacement is calculated, a previously defined number of the closest systems are selected and the matrices and vectors are weighted by distance. The corresponding equation is given by:

$$\tilde{w}_n(q_{\text{I}}) = \frac{1}{N\,\|q_{\text{I}} - q_n\|^2} \tag{4}$$

$$N = \sum_{n=1}^{n_a} \tilde{w}_n(q_{\text{I}}). \tag{5}$$

Here, N is the normalizing factor, q_I is the interpolation point and q_n are the order reduced training points. The value n_a gives the number of *active systems* – this is the number of snap shots used for the interpolation. By applying the w factors directly to the coordinates the simulation becomes faster and more stable.

Another possibility is the usage of a linear, vectorial interpolation. Here, the linear independance of the vectors used to interpolate has to be ensured to get a unique solution. Hence, the amount of interpolation vectors has to be smaller than the amount of reduced DOF. The equations are

$$\tilde{w}_1(q_I) = 1 - \sum_{n=2}^{n_a} a_n, \qquad (6)$$

$$\tilde{w}_n(q_I) = a_n, \qquad\qquad\qquad n \neq 1, \qquad (7)$$

where the a_n can be calculated with the following equation system:

$$q_I = q_1 + \sum_{n=2}^{n_a} a_n(q_n - q_1). \qquad (8)$$

The vectorial interpolation concept is slower in calculation time but negative weights are feasible. The latter gives an advantage if the training data is not sufficiently covering the simulated trajectory.

Even with the usage of the approximation given in equation (3), the interpolation approaches lead to a nonlinear problem because weight functions or vector prefactors have to be calculated. Consequently, a Newton-Raphson iteration is needed.

Alternatively, it is possible to implement a prediction, which is a two iteration approach (see Fig. 2): The internal load of the current displacements q_n is determined by interpolation, the time integration is performed and the displacements q_{temp} are

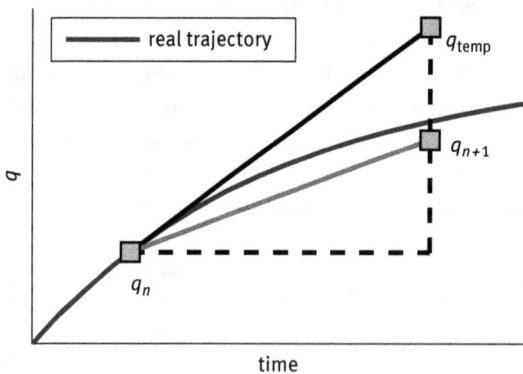

Fig. 2. Scheme of a predictor based time integration. The mean of the stiffnesses at step q_n and q_{temp} leads to a more realistic trajectory (red).

FEM Training simulation with data export

↓ coefficients, displacements ↓

calculate int. load from data export for all snapshots x_i

↓ $K_{lin}(x_i), f_{lin}(x_i)$ ↓

determine reduction matrices S from snapshot trajectory

↓ S ↓

store MOR int. forces of snapshot points	apply (any) time dependent load

↓ $S, K_{lin}^{red}(x_i), f_{lin}^{red}(x_i)$ ↓

calculate new trajectories by interpolating the MOR data	in time step $n+1$ interpolate $K_{lin}(x_n)$, $f_{lin}(x_n)$ from training data
	use $K_{lin}(x_i), f_{lin}(x_i)$ to copute x_{temp} with any time integration method
	determine $K_{lin}(x_{temp}), f_{lin}(x_{temp})$
	use the mean of the two int. loads to obtain x_{n+1}

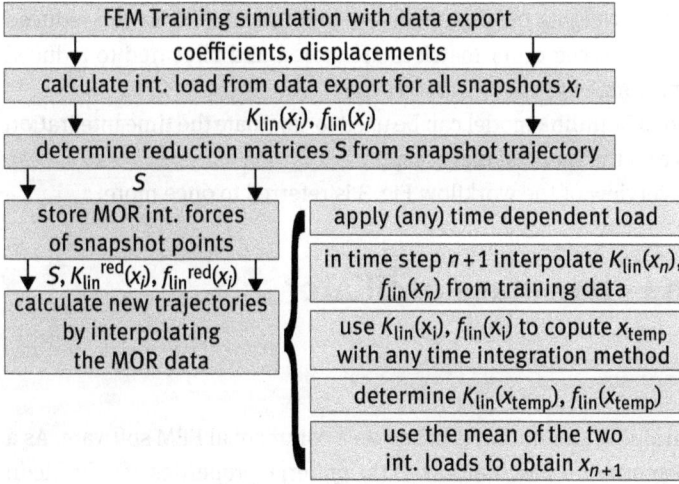

Fig. 3. Flow chart of the workflow and the time integration with the predictor approach.

obtained. Now, the internal forces at q_{temp} are interpolated and the time integration is computed once more by starting again at q_n. This time, the mean of the two derived internal loads is used.

In the example shown in Fig. 2 the stiffness at x_n is too small, whereas it is too big at x_{temp}, making the mean a valid approximation. Another illustration of the predictor approach can be found in the flow chart in Fig. 3. Here, the full implementation workflow of a TPWL concept, as described in the next subsection, is shown.

2.4 Workflow and implementation for MEMS gyroscopes

To gather a sufficient amount of interpolation data, a transient nonlinear FEM simulation has to be run first. Consequently, an FEM model has to be generated and time-dependant forces need to be applied to obtain all relevant sensor deformations. Possible points of force application are the nodes at the electrode positions. An FEM training simulation with similar parameter settings as later used in the MOR is reasonable.

The $f_{lin}(x_n)$, $K_{lin}(x_n)$ from Eqs. 2, 3 often cannot be addressed explicitly during the solution process of commercial FEM tools. Instead, the coefficients of the solved equations can be exported. As a consequence, the output depends on the analysis type as well as the used solver.

In the next step, the internal load information has to be derived from the stored data - Matlab or similar tools can be used for this computation. If a POD is used, the reduction matrices S can be calculated from the training trajectory. In this case, the singular values also give weights on the necessary amount of DOF in the reduced system. Afterwards, the reduced order matrices are stored and allocated to reduced or full displacement vectors.

Finally, a Matlab or a Simulink model can be used to calculate the time integration with new parameters and the reduced matrices.

For a flow chart overview of the workflow Fig. 3 is referred to once more.

3 Application I - one mass oscillator

3.1 Setup

The following TPWL training simulation is based on a commercial FEM software. As a test vehicle, a nonlinear one mass oscillator with the material properties of polysilicon as shown in Fig. 4a is used. To make the approach realistic, perforated mass and material (i. e. polysilicon) are similar to the ones used in common MEMS gyroscopes as shown in e. g. [1]. A strong nonlinear oscillation is achieved, since the substrate connections behave like double clamped beams. The mass has a size of $400\,\mu m \times 400\,\mu m$ and is meshed with a total of about 1700 nodes. A $Q > 1000$ is chosen.

In the training run, the amplification of the oscillator needs to be significantly larger than in the subsequent MOR simulation. Otherwise, there wouldn't be enough sample points at high deflections for the predictor approach to work properly (see also Fig. 2). The attempts show that a sine actuation force amplitude of $5\,\mu N$ along the first eigenmode, driven with the corresponding resonance frequency, leads to a clear nonlinear behavior after very few oscillation periods. Therefore, this setting is used in the MOR simulation. In the training run, the force amplitude is doubled to $10\,\mu N$. In addition, the second eigenmode is actuated with the same force but driven with the second resonance frequency (see Figs. 4b, 4c). This second force is not applied in the MOR simulation but it makes the training trajectory more complex and therefore, gives this setup a more realistic test for the TPWL approach. The forces are applied to the central point of the perforated mass. 200 load steps are computed at a sampling rate of 40 time intervals per period of the first eigenfrequency. This is equal to 5 sine periods – other settings for the training run have been tried (see 3.2).

A second simulation with the FEM tool is performed to get a reference value. Now, 10 000 load steps are computed, the amplitude of the force is changed to $5\,\mu N$ and instead of actuating two eigenmodes, a force just along the first mode is applied. 250 oscillation periods are evaluated in total. The increase of the load step amount gives the possibility to verify the applicability of the TPWL method further away from the training run scenario.

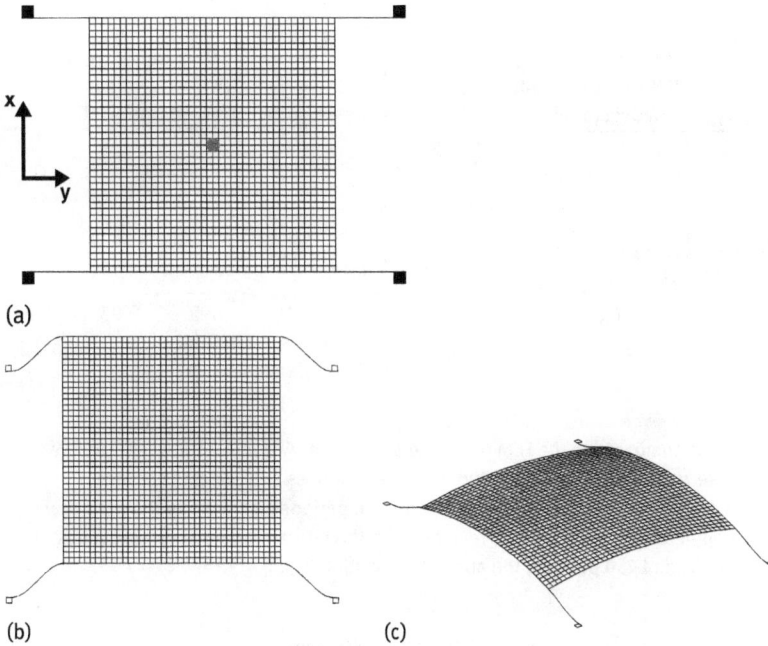

(a)

(b)   (c)

Fig. 4. (a) The simulated geometry with (red) force application point and (blue) substrate bonds, along with (b) the first and (c) the second eigenmodes are shown. The first mode is in-plane in x-direction, the second mode is out-of-plane in z-direction.

A Matlab implementation is used to simulate the time integration of the 10 000 load steps with the MOR approach. The system is reduced to 5 DOF[1], whereas n_a is 2. The same time integration scheme as in the FEM tool is used.

3.2 Results

Figure 5 compares the MOR results to the FEM tool reference value. In Fig. 5a the first 25 and in Fig. 5b one of the last of the 10 000 oscillation periods are shown. Here, it is important to keep in mind that the snapshot points also have a displacement in z-direction, whereas the movement patterns in the Matlab simulation are just at $z = 0$ for all nodes.

---

1 In [28] "'2 DOF'" was stated incorrectly.

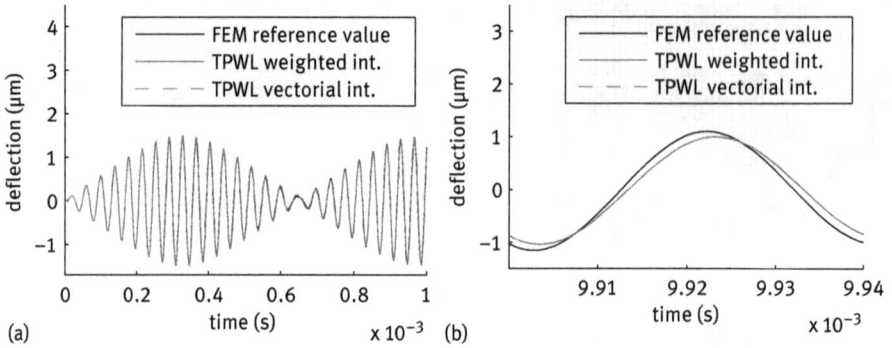

Fig. 5. MOR approaches compared to the FEM reference simulation. Two interpolation procedures were tested, a positive weight function calculated from the distances and a vectorial concept - both leading to very similar results. (a) The first 25 oscillation periods clearly show the nonlinear behavior - the MOR simulations are in good agreement with the reference trajectory. (b) Even after nearly 10 000 load steps and 250 periods, the absolute deviation is just 4.98 %.

The two interpolation methods given in the last subsection are used in the MOR simulation. It can be seen from the Figs. 5a, 5b that both lead to a similar result in case of the test device. The results hardly change if n_a is set to 1, which essentially deactivates the interpolation and uses the closest training point.

The absolute error of the simulation is only 4.98 % for the amplitude maximum in the last 10 oscillation periods. The phase shift is less than the phase resolution given by the sampling rate (9°). This is an excellent match and the precision is hardly limited by the MOR itself – the accuracy can be increased by adding snapshots. Calculation times are reduced from several hours for the reference simulation to about 15 seconds – this is a factor of more than 10^3.

As explained in 2.2, the interpolation concept itself can decrease the simulation time even if used with full system matrices. Using the model above, a time decrease by a factor of 10 can be achieved.

The graph in Fig. 5 shows the typical nonlinear oscillation characteristics: Due to stress stiffening effects the resonance frequency is changing with increasing displacements and the drive turns off-resonant - the amplitude is decreasing before it starts to increase again. A linear simulation approach gives a very different qualitative behavior. Due to the small damping, the oscillation amplitude is increasing with each period to unrealistically high amounts.

Other attempts show that load step amount and sampling rate of the training run shouldn't be much smaller as the results become less accurate.

4 Application II - nonlinear gyroscope

4.1 Setup

In [29] an especially nonlinear designed MEMS gyroscope has been introduced. The structure, linear detection and actuation mode of the sensor are shown in Figs. 6. Here, the comb electrodes that actuate the sensor can be seen in as well.

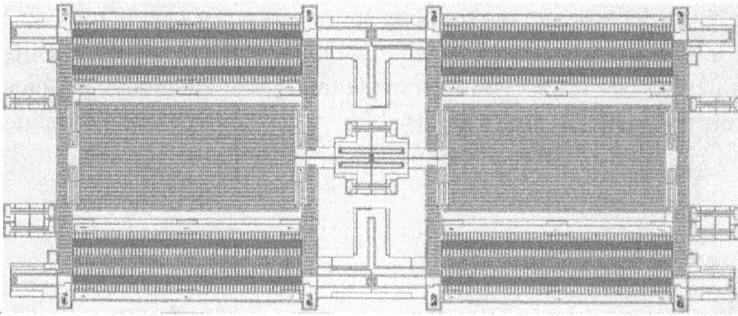

(a)

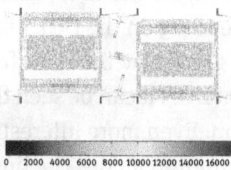

(b) Actuation mode at 24398 Hz

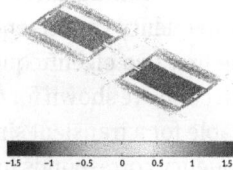

(c) Detection mode at 27415 Hz

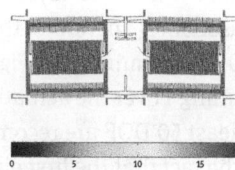

(d) Coupling spring mode at 52285 Hz

Fig. 6. (a) The geometry of the nonlinear designed MEMS gyroscope, (b) the linear actuation mode, (c) the linear detection mode are shown and (d) the coupling spring mode are shown.

The goal of this section is to examine the influence of different training runs on the MOR simulation. In the test calculations the mode of the coupling spring shown in Fig. 6d shall be excited and its eigenfrequency in dependence of the actuation amplitude will be determined with a fast Fourier transform (FFT). Therefore, three trainings based on a finite element model with altogether 18684 nodes are performed.

- Training 1 also is used to test the dependency of the actuation deflection on the eigenfrequencies of the system. Therefore, only the actuation mode is actuated.
- In training 2 the actuation mode and the coupling spring mode are excited at the same time.
- In training 3 again actuation mode and coupling spring mode are excited, but now the actuation is separated: The actuation mode is excited in the first two thirds of the training run. After reaching a certain amplitude the force is removed and the coupling spring is actuated.

In all three training simulations the actuation mode reaches an amplitude of about $20\,\mu m$. The maximum deflection of the coupling spring mode is $4.5\,\mu m$ in training 2 and $7.5\,\mu m$ in training 3. In each training 490 snap shots are generated – the sampling rate is 70 steps per periode of the actuation mode (altogether 7 periods). As force application points the comb electrodes are used for the actuation. For the coupling spring deflection a force on all nodes of the spring is applied.

4.2 Analyzing the MOR data

4.2.1 Convergence

Before a transient system simulation can be executed a study of the MOR data is reasonable. Therefore, the eigenfrequency convergence depending on the DOF of the MOR is examined. In Figs. 7 the first two eigenfrequencies gained by the data from training 1 over the actuation deflection are shown for different DOF. It can be seen that at least 50 DOF are recommendable for a transient simulation. Even more interesting is the fact that the first eigenfrequency corresponds to the actuation mode. The linear modal analysis assigns frequency number 3 to this mode. The other frequency cannot be identified with a linear eigenmode. The reason for this is the statistical behavior of the POD which creates mode shapes and frequencies corresponding to the training run.

4.2.2 Mode shapes

As outlined above the POD mode shapes correspond to the applied training run and do not match the linear mode shapes from the modal analysis. Consequently, it is interesting to have a close look on these results. The first eigenmode in all trainings can be identified with the linear actuation mode (mode number 3) – the frequencies are identical. Higher modes in training 1 do not correspond to any other physical modes and are all in-plane. The second mode shape in training 2 is equal to the coupling spring mode, which is mode number 7 in the linear analysis.

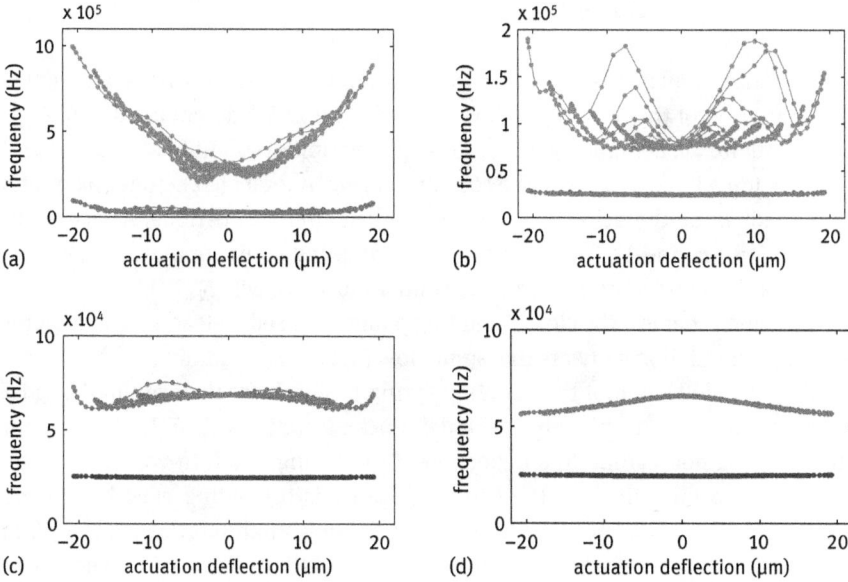

Fig. 7. The first two eigenfrequencies gained by the data of training 1 over the actuation deflection are shown. The dependency of the frequencies from the DOF can be seen by the different graphics.

Though, the eigenfrequencies are different (linear: 52285 Hz, training 2: 44195 Hz). Unexpected is the shape of mode number 2 in training 3. Here, an identification with the linear mode 7 is not possible. Instead, the coupling spring deflection is similar to a superposition of modes 2 and 3 (see Figs. 8). All other modes in training 2 and 3 are unphyical.

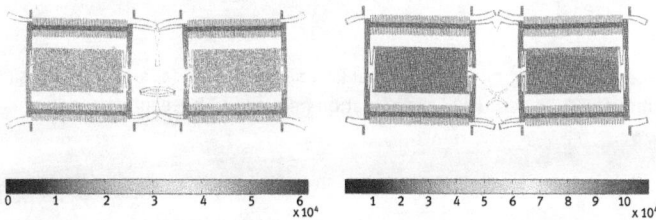

(a) POD mode 2 at 48090 Hz   (b) POD mode 3 at 55148 Hz

Fig. 8. The second (a) and the third (b) POD eigenmode of training number 3 are shown.

4.3 Transient simulations

A test simulation shall be executed with the data of all three training runs. The goal is the determination of the coupling spring frequency for two different actuation amplitudes. Therefore, an actuation amplitude is applied (initial conditions are used) and a delta-peak force is used to excite the coupling spring. A total of 22 actuation periods are simulated, while the delta-peak force is applied after 2.25 periods. As actuation amplitudes 6.8 μm and 13.5 μm are used – the coupling spring frequency has been measured for these two values in [29]. The simulation is run with 50 DOF and without an interpolation scheme – the closest training points are used instead and a predictor is applied. As simulation software the signal flow tool Matlab Simulink is used.

In Fig. 9 the deflection of the coupling spring is shown for the trainings 1 and 2 (13.5 μm amplitude) – the moment of the delta-peak force is marked. It can be seen, that training 1 is not leading to a deflection of the spring at all. In contrast, in the simulation based on training 2 the coupling spring is oscillating even before the moment of the delta-peak force. The change in oscillation amplitude due to the force is a lot smaller than expected and the qualitative shape of the oscillation does not seem to be stable.

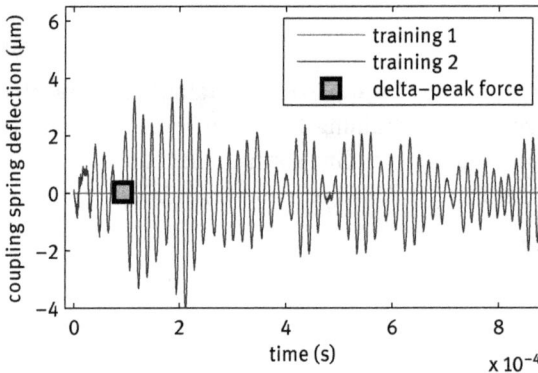

Fig. 9. The oscillation of the coupling spring during the signal flow simulation. A delta-peak force is applied on the coupling spring after 2.25 periods of the actuation frequency. The results based on training runs 1 and 2 are shown.

The explanation for this result is that most nonlinear data gained from training 2 contain actuation and coupling spring deflections at the same time, and therefore a coupling between these two modes is given and energy transfers occur. On the other hand, in training 1 the POD generates no mode describing the coupling spring oscillation at all, and consequently an actuation of the spring does not happen.

The transient simulation based on training 3 with an actuation amplitude of 13.5 μm is shown in Fig. 10. As now there are snap shots without a coupling spring excitation, the oscillation of the spring is much smaller before the delta-peak force is applied than in the previous approach. The qualitative shape of the oscillation after the force application is also more stable now. An even better result might be achieved by a more advanced training or in general with more snap shots.

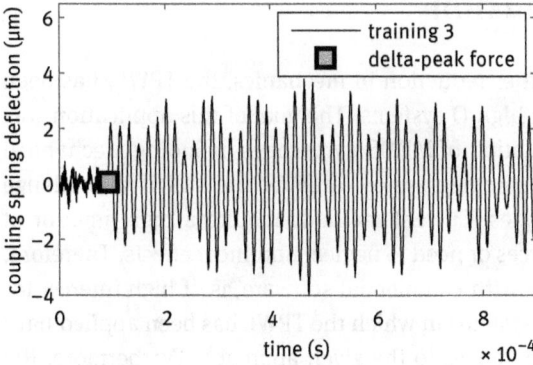

Fig. 10. The oscillation of the coupling spring during the signal flow simulation. A delta-peak force is applied on the coupling spring after 2.25 periods of the actuation frequency. The result based on training run 3 is shown.

Tab. 1. Frequency shift of the coupling spring mode from the frequency at zero actuation amplitude.

method	frequency shift at 6.8 μm act. amp.	frequency shift at 13.5 μm act. amp.
measurement	+1160 Hz	+3981 Hz
FEM	+1089 Hz	+3684 Hz
TPWL (training 2)	+1191 Hz	+3574 Hz
TPWL (training 3)	+1191 Hz	+2383 Hz

To compare the frequency shift of the coupling spring mode from the frequency at zero actuation amplitude, an FFT is applied on the result data. In table 1 the simulation results are compared to the measurements from [29]. Furthermore, results of a full transient FEM calculation performed in [29] are added. All results are in good agreement with the measurements but interestingly, the quantitative values for the change in frequency compared to the measurements are better for training 2 than for training 3. The interpretation is that in training 3 the coupling spring mode is only actuated during the last third of the calculation time. Consequently, there are less snap shots with a deflection of the coupling spring than in training 2 (in both trainings 490 steps are calculated). Therefore, it can be assumed that a qualitative and quantitative good result can be achieved by a training simulation based on training 3 with more snap shots. Anyway, the quantitative results based on training 2 are comparable to the results from the full FEM simulation which is a great success for such a complex system.

5 Summary and conclusion

A nonlinear approach of model order reduction in mechanics, the TPWL, has been introduced with a focus on MEMS high Q systems. The goal of this application was the time efficient full system simulation of MEMS gyroscopes including mechanical nonlinearities, since their highly complex system architecture still is a challenge for transient computation techniques. Current methods require a high amount of calculation time and system resources or need to neglect nonlinear effects. Therefore, a novel simulation process, usable with commercial software, is of high interest for the industrial development. Other systems in which the TPWL has been applied have been introduced to outline the differences to the given approach. Furthermore, the nonlinear expansion of the MOR, introduced in [20], has been compared to the TPWL in a qualitative discussion.

To make the TPWL approach more understandable, first the implementation of the method with training data from commercial software has been shown schematically. The mathematical fundamentals were outlined and a predictor method to increase the accuracy of the procedure was introduced.

The first approach is a general verification of the methodology. Therefore, a simple MEMS test model has been used to show that the TPWL can decrease the simulation time by a factor of 10^3 from hours to seconds with hardly any loss in precision – compared to a FEM calculation the discrepancy is only 4.98 %.

In a second application, a more complex system was used – an especially nonlinear designed MEMS gyroscope. This setup is more similar to systems which are under current development. Therefore, the analysis of the training influence on POD mode shapes and eigenfrequencies gave a good estimation for the applicability of the methodology in the industrial development process and on how the POD modes should be interpreted. The discussion on the frequency convergences in dependance of the system order gave important hints for the setup of a fulll MEMS gyroscope MOR simulation. Eventually, a nonlinear MOR signal flow simulation at which the calculated results and the measurements are in a good agreement could be performed.

Further works will include more studies on appropriate training trajectories in MEMS gyroscopes, as well as further implementations of the TPWL approach in other MEMS gyroscope systems. Moreover, a direct comparison to the method in [20] is necessary in the context of MEMS gyroscopes.

Acknowledgment: The authors would like to thank Rolf Scheben for his continued support of this work and all the colleagues at Robert Bosch GmbH for many helpful discussions and comments.

Bibliography

[1] R. Neul, U.-M. Gomez, K. Kehr, W. Bauer, J. Classen, C. Döring, E. Esch, S. Gotz, J. Hauer, B. Kuhlmann, C. Lang, M. Veith und R. Willig, "Micromachined angular rate sensors for automotive applications", *IEEE Sensors Journal*, Bd. 7, Nr. 2, S. 302–309, Feb. 2007.

[2] J. Marek, "MEMS for automotive and consumer electronics", *Solid-State Circuits Conf. Digest of Technical Papers (ISSCC), 2010 IEEE International*, Feb 2010, S. 9–17.

[3] V. Kempe, *Inertial MEMS : Principles and Practice*, 1. Auflage. Cambridge, UK: Cambridge university press, 2011.

[4] C. C. Painter und A. M. Shkel, "Structural and thermal analysis of a mems angular gyroscope", S. 86–94, 2001. [Online]. Unter: http://dx.doi.org/10.1117/12.436630

[5] A. Ferreira und S. Aphale, "A survey of modeling and control techniques for micro- and nanoelectromechanical systems", *Systems, Man, and Cybernetics, Part C: Applications and Reviews, IEEE Trans. on*, Bd. 41, Nr. 3, S. 350–364, May 2011.

[6] V. Kolchuzhin, M. Naumann und J. Mehner, "Recent developments in reduced order modeling based on mode superposition technique", *EuroSimE*, Linz, Austria, Apr. 2011, S. 252–257.

[7] D. Gugel, W. Dötzel, T. Ohms und J. Hauer, "Reduced order modeling in industrial MEMS design processes", *Eurosensors XX*, Göteborg, Sweden, Sep. 2006, S. 48–55.

[8] G. Martin und G. Smith, "High-level synthesis: Past, present, and future", *Design Test of Computers, IEEE*, Bd. 26, Nr. 4, S. 18–25, July 2009.

[9] W. Xue, J. Wang und T. Cui, "Modeling and design of polymer-based tunneling accelerometers by ansys/matlab", *Mechatronics, IEEE/ASME Trans. on*, Bd. 10, Nr. 4, S. 468–472, Aug 2005.

[10] G. Duffing, *Erzwungene Schwingungen bei veränderlicher Eigenfrequenz und ihre Technische Bedeutung*. Braunschweig, Germany: Vieweg, 1918.

[11] H. K. Lee, J. C. Salvia, S. Yoneoka, Y. Q. Qu, R. Melamud, S. Chandorkar, M. A. Hopcroft, B. Kim und T. W. Kenny, "Stable oscillation of mems resonators beyond the critical bifurcation point", *Hilton Head Workshop*, Hilton Head Island, USA, Jun. 2010, S. 70–73.

[12] D. Gugel, "Ordnungsreduktion in der mikrosystemtechnik", Dissertation, Technische Universität Chemnitz, Chemnitz, Germany, 2008.

[13] V. Kaajakari, T. Mattila, A. Lipsanen und A. Oja, "Nonlinear mechanical effects in silicon longitudinal mode beam resonators", *Sensors and Actuators A: Physical*, Bd. 120, Nr. 1, S. 64 – 70, 2005. [Online]. Unter: http://www.sciencedirect.com/science/article/pii/S092442470400826X

[14] M. N. Albunni, V. Rischmuller, T. Fritzsche und B. Lohmann, "Model-order reduction of moving nonlinear electromagnetic devices", *IEEE Trans. on Magnetics*, Bd. 44, Nr. 7, S. 1822–1829, Jul. 2008.

[15] A. Verhoeven, T. Voss, P. Astrid, E. ter Maten und T. Bechtold, "Model order reduction for nonlinear problems in circuit simulation", S. 1 021 603–1 021 604, Dec. 2007.

[16] J. He und L. Durlofsky, "Constraint reduction procedures for reduced-order subsurface flow models based on POD-TPWL", *Int. J. Numer. Meth. Engng.*, Bd. 103, S. 1–30, Jul. 2015.

[17] M. Rewieński und J. White, "A trajectory piecewise-linear approach to model order reduction and fast simulation of nonlinear circuits and micromachined devices", *IEEE/ACM Iccad*, San José, USA, Nov. 2001, S. 252–257.

[18] M. Rewieński und J. White, "Model order reduction for nonlinear dynamical systems based on trajectory piecewise-linear approximations", *Linear Algebra and its Applications*, Bd. 415, Nr. 2–3, S. 426–454, 2006, special Issue on Order Reduction of Large-Scale Systems. [Online]. Unter: http://www.sciencedirect.com/science/article/pii/S0024379504001727

[19] D. Agrawal, J. Woodhouse und A. Seshia, "Modeling nonlinearities in mems oscillators", *Ultrasonics, Ferroelectrics, and Frequency Control, IEEE Trans. on*, Bd. 60, Nr. 8, S. 1646–1659, Aug. 2013.

[20] F. Bennini, J. Mehner und W. Dötzel, "Computational methods for reduced order modeling of coupled domain simulations", *Transducers*, Munich, Germany, Jun. 2001, S. 260–263.

[21] J. Mehner, *Modal-Superposition-Based Nonlinear Model Order Reduction for MEMS Gyroscopes*. Wiley-VCH Verlag GmbH & Co. KGaA, 2013, S. 291–309. [Online]. Unter: http://dx.doi.org/10.1002/9783527647132.ch12

[22] M. Dienel, M. Naumann, A. Sorger, D. Tenholte, S. Voigt und J. Mehner, "Stable oscillation of mems resonators beyond the critical bifurcation point", *Vacuum, Special Issue: Sensors*, Bd. 86, Nr. 5, S. 536–546, Jan. 2012.

[23] J. M. Ortega und W. C. Rheinboldt, *Iterative Solution of Nonlinear Equations in Several Variables*, Serie Classics in Applied Mathematics. Philadelphia, USA: SIAM, 1987, Nr. 30.

[24] P. Wriggers, *Nonlinear Finite Element Methods*. Berlin, Germany: Springer, 2008.

[25] T. Bechtold, M. Striebel, K. Mohaghegh und E. t. Maten, "Nonlinear model order reduction in nanoelectronics: Combination of pod and tpwl", *Appl. Math. Mech.*, Bd. 8, Nr. 1, Bremen, Germany, Dec. 2008, S. 10 057–10 060.

[26] I. T. Jolliffe, *Principal Component Analysis*, 2. Auflage. Berlin, Germany: Springer, 2002.

[27] D. Vasilyev, M. Rewienski und J. White, "A tbr-based trajectory piecewise-linear algorithm for generating accurate low-order models for nonlinear analog circuits and mems", *Design Automation Conf., 2003. Proceedings*, June 2003, S. 490–495.

[28] M. Dorwarth, S. Kehrberg, R. Maul, R. Eid, F. Lang, B. Schmidt und J. Mehner, "Nonlinear model order reduction for high Q MEMS gyroscopes", *Multi-Conf. on Systems, Signals Devices (SSD), 2014 11th International*, Castelldefels, Spain, Feb 2014, S. 1–4.

[29] S. Kehrberg, "Charakterisierung und Simulation der Nichtidealitäten mikromechanischer Drehratesensoren", dissertation, Technische Universität Chemnitz, Chemnitz, Germany, 2015.

Biographies

Markus Dorwarth Markus Dorwarth was born 1986 in Karlsruhe, Germany. He studied physics at the Stuttgart University of Technology and the National University of Signapore. He received his diploma (Dipl.-Phys.) in 2011. Since 2012 he is working as doctoral candidate at the Robert Bosch GmbH in Stuttgart, Germany in the field of microsystem technologies with a focus on MEMS gyroscopes. The work project on simulation methodologies is a collaboration with the professorship of microsystems and medical engineering at the Chemnitz University of Technology.

Steven Kehrberg Steven Kehrberg received his Dipl.-Ing. degree in electrical engineering with focus on microsystems and precision engineering at the Chemnitz University of Technology, Germany, in 2011. Since 2011 he has been working at the Automotive Electronics department of Robert Bosch GmbH in Reutlingen, Germany. His research interests include design, simulation and characterization of MEMS gyroscopes.

Robert Maul Robert Maul studied Physics in Jena and obtained his degree in 2006. After furthering his studies, he received his Ph.D. in Physics from the Karlsruhe Institute of Technology (KIT) in 2010. His research at the Institute of Nanotechnology (INT) and Steinbuch Center for Computing (SCC) was on quantum transport in nanostructures, i. e. simulation of charge transfer in dynamic quantum systems. After having joined the corporate research department of the Robert Bosch GmbH in Gerlingen he started working in MEMS design and related simulation method development. Since 2015 he is working as designer of future MEMS gyroscopes for consumer applications in the development department of the Robert Bosch GmbH in Reutlingen.

Rudy Eid Rudy Eid studied Electrical Engineering at the Notre Dame University, Zouk Mosbeh, Lebanon, where he graduated in 2003. He obtained his M.Sc. in Information and Automation Engineering from the University of Bremen in 2005, and his Ph.D. in Control Theory from the Technische Universität München, Germany, in 2009, where he then pursued his postdoctoral research. In 2011, he joined the corporate research department of the Robert Bosch GmbH in Gerlingen, Germany. Since 2014, he is a MEMS designer within the Automotive Electronics division of the Robert Bosch GmbH in Reutlingen, Germany. His current research areas of interest are the design and simulation of MEMS inertial sensors and model order reduction of dynamical systems.

Florian Lang Born in 1980 in Innsbruck, Austria, studied experimental physics at University Innsbruck. Master thesis at Uppsala University, Sweden in Experimental Particle Physics, PhD thesis at University Innsbruck on Ultracold Quantum Gases. Works for Robert Bosch GmbH since 2011.

Benjamin Schmidt Benjamin Schmidt was born in Neumünster, Germany in 1977. He received his Diplom degree in physics from RWTH Aachen in 2004 and his Dr. rer. nat. degree from the Karlsruhe Institute of Technology in 2008. From 2008 till 2014 he was working for the corporate research department of Robert Bosch GmbH, Gerlingen, Germany focussing on the design of and methodology development for micro electro mechanical gyroscopes. He is now working with the starter motors and generators devision of Robert Bosch GmbH.

Jan Mehner Jan E. Mehner is Professor for Microsystems and Medical Engineering at the Chemnitz University of Technology in Germany. He received a Dr.-Ing. and Dr.-Ing. habil degree in Electrical Engineering and Information Technology from the same University in 1994 and 1999, respectively. From 1998 to 1999 he was a Visiting Scientist at the Massachusetts Institute of Technology in the field of MEMS design. In 2013 he became member of the Academy of Science in Saxony, Germany. His research interests include analytical and numerical methods to design microsystems, CAD-tools and computational algorithms for coupled physical domains and model order reduction.

M. Meister, U. Liebold, A. Jäger, S. Thiele, R. Weirauch, D. Gäbler
and K. Bach

Monolithic Integrated CMOS Ambient Light Sensor

Abstract: Ambient light sensors are often used in electronic devices for brightness control. Therefore, the luminosity function of the human eye has to be adapted by a semiconductor sensor. This is often reached by applying an optical filter on the sensor surface. This paper describes a ambient light sensor, which can be produced without an optical filter on the surface. It can be realized by a photodiode stack of 3 pn-junctions and the linear combination of the resulting photo-generated currents. This concept is verified by numerical simulations and measurements.

Keywords: Ambient light sensor, photodiode, photodiode stack, luminosity function, photo diode sensitivity, depth of light penetration, beer-lambert law.

1 Introduction

Ambient light sensors are usually realized as an integrated silicon photodiode with an optical coating on top. The luminosity function is given by the optical parameters of coating. This optical filtering requires a post-processing after the semiconductor process. Hence, it needs extra time and charges. This paper describes a possibility to realize an ambient light sensor using a standard $0.35\,\mu$m-CMOS-process. For realization of this sensor only existing layers of the underlying CMOS-process were used. In this way no post-processing after semiconductor process is necessary.

The presented ambient light sensor based on different penetration depths of incident light in silicon. This effect can be used to separate charge carriers from different light wavelengths within a photodiode stack of 3 pn-junctions. The resulting photo-generated currents were linear combined by the integrated circuitry. So the sensitivity characteristic can be adapted to the luminosity function of the human eye. This behavior was verified by numerical simulations and measurements.

M. Meister, U. Liebold and A. Jäger: Institute for Microelectronics and Mechatronics Systems, 98693 Ilmenau, Germany, e-mails: michael.meister@imms.de, ulrich.liebold@imms.de, andre.jaeger@imms.de.
S. Thiele, R. Weirauch, D. Gäbler and K. Bach: X-FAB Semiconductor Foundries AG, 99097 Erfurt, Germany, e-mails: sebastian.thiele@xfab.com, robin.weirauch@xfab.com, daniel.gaebler@xfab.com, konrad.bach@xfab.com.

De Gruyter Oldenbourg, ASSD – Advances in Systems, Signals and Devices, Volume 6, 2018, pp. 217–226.
https://doi.org/10.1515/9783110448375-014

2 Structure and idea of operation

Photodiodes are used to change light into an electrical current. During this process incident photons generate charge carriers, which will be separated by the electric field of a pn-junction. The ratio between generated photo current and incident light power is called sensitivity and has a strong dependence on the light's wavelength. Silicon photodiodes have their maximum sensitivity within the infrared wavelength range (Figure 1). In contrast the luminosity function of human eye has its maximum at $555\,nm$ for photopic vision and $507\,nm$ for scotopic vision.

Normalized Sensitivity of human eye and silicon photodiode

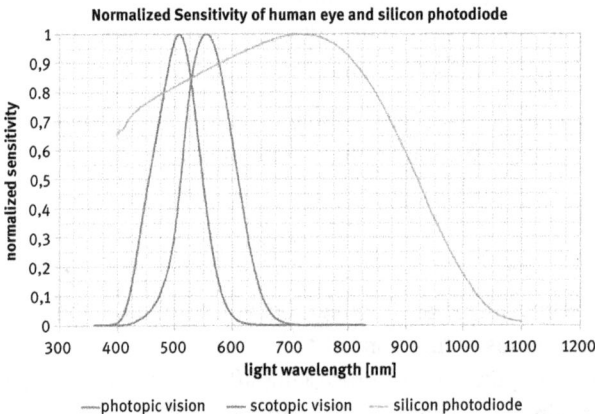

Fig. 1. Normalized Sensitivity of human eye and silicon photodiode.

Sensitivity characteristics of human eye and silicon photodiode are very different. Especially, the infrared part of a silicon photodiode has to be attenuated. This can be done by separating the photo-generated currents of different incident light wavelengths. Therefore, the penetration depth of light in silicon can be used. The penetration depth of photons into silicon depends on the wavelength of incident light. Blue light at $400\,nm$ only has a penetration depth of less than $1\,\mu m$ into silicon. In contrast to that red light at $630\,nm$ will be completely absorbed at about $17\,\mu m$ depth.

In Figure 2 the depth of light penetration vs. wavelength λ is shown for four different values of absorption A. The green line represents the penetration depth where 99% of incident light is absorbed. The rate of absorption is defined by the absorption coefficient α, where the extinction coefficient n'' is the imaginary part of the refractive index.

$$\alpha = -\frac{4\pi}{\lambda}n'' \tag{1}$$

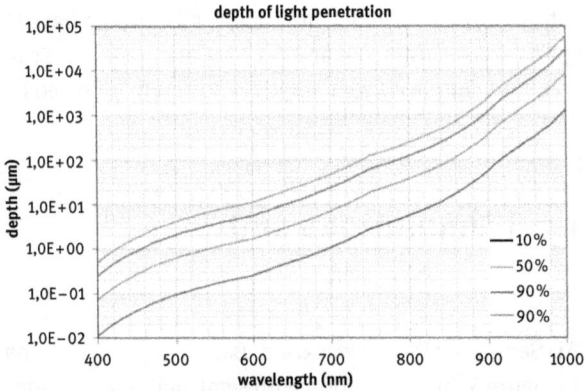

Fig. 2. Depth of light penetration in silicon.

Due to the Beer-Lambert law [7] $T = e^{-\alpha \cdot d}$ the penetration depth d is given by:

$$d = -\frac{\lambda \ln(1-A)}{4\pi \cdot n''} \tag{2}$$

For light at a wavelength of $630\,nm$ an extinction coefficient $n'' = 0.0136$ [3] is given. Using (1) an absorption coefficient $\alpha = -271.37 \cdot 10^{-6}\,nm^{-1}$ can be calculated. This results to a penetration depth $d = 16.98\mu\,m$, where 99% of incident light were absorbed.

3 Structure and idea of operation

This effect can be used to detect photo-generated charge carriers in different depths of silicon. Therefore stacked pn-junctions within a single semiconductor device could be applied for it. For realization of this structure only available standard layers of existing process should be used. The given technology allows a stack of 3 pn-junctions. The first pn-junction (contacts c1 and a1) has to be as close as possible to the surface of silicon. It is represented by a highly doped n+ implant within low doped p- well. Here, the sensitivity for light of short wavelengths is very high. First pn-junction is placed within a very low doped deep n-well. So a second pn-junction (contacts a1 and c2) was created, which allows detection of mid-range light wavelengths. Third pn-junction (contacts c2 and a2) is built between deep n-well and p-substrate.

The described structure allows the separation of the photo currents, which are generated from different light wavelengths. For each pn-junction a single optical sensitivity characteristic can be determined. These characteristics can be combined by linear combination to a required sensitivity characteristic. Linear combination means multiplying each current by a specific coefficient. Finally, all currents have to be added. Hence, the infrared part can be subtracted and the blue and green part can be amplified. This can be realized by a simple current amplifier structure followed by a current summation.

4 Results of numerical simulations and measurements

For numerical simulations TCAD Sentaurus Device was used. Base for all simulations is the device structure shown in Figure 3. In cooperation with semiconductor foundry X-FAB ideal dopant profiles were adapted to measurement results. Light incident und generated photo current in each single pn-junction could be simulated.

A test structure of a stacked photodiode was processed and characterized. Thereby, a monochromatic light was directed to the stacked photodiode and all resulting photo currents were measured. This process was executed within a light wavelength range from $400\,nm$ to $1100\,nm$ in steps of $5\,nm$.

The measurement result is shown in Figure 4 as solid lines for all 4 available contacts c1, a1, c2 and a2. Dashed lines show the corresponding simulation results. Contact c1 is connected only to n+-doped region as close as possible to the surface. Hence, the sensitivity response has its maximum at blue light. Contact a1 is connected to the first and the second pn-junction. Here, charge carriers from both pn-junctions can be detected. Due to this behavior light wavelength for the maximum of sensitivity response is shifted to the green light range. Both, simulation and measurement results, show similar characteristics. Contacts connected to deeper pn-junctions have their sensitivity response maximum at longer light wavelengths.

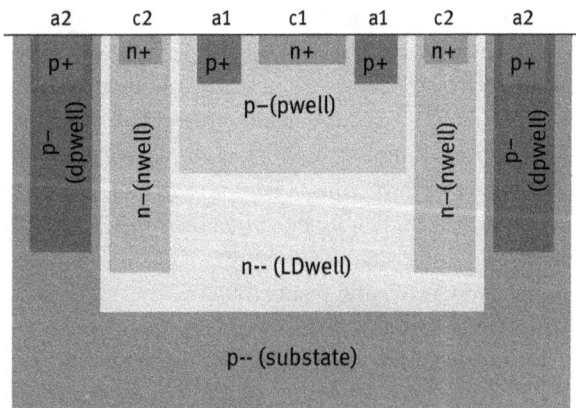

Fig. 3. Cross section of stacked photodiode.

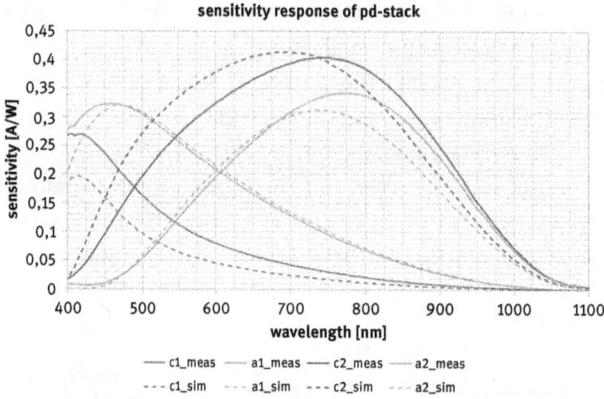

Fig. 4. Simulation and measurement results of sensitivity response of stacked photodiode.

Based on the measurement results linear combination can be used to adapt the resulting sensitivity response to luminosity function. Each current was multiplied by a specific coefficient and finally all weighted currents were added. Method of least squares was applied to solve this fitting problem. The resulting coefficients for this example were:

coefficient for photo current of contact c1	=	-14
coefficient for photo current of contact a1	=	14
coefficient for photo current of contact c2	=	-1
coefficient for photo current of contact a2	=	0

The calculated sensitivity response of the stacked photo diode is shown in Figure 5. In comparison to Figure 1 the sensitivity response has a good adaption to luminosity function of the human eye.

5 Integrated circuitry

The shown concept was used to realize an integrated ambient light sensor, which consists of the stacked photo diode as the optical sensor element and an integrated circuitry to implement the linear combinations of the photo-generated currents. This structure can be separated into four parts. At the beginning the sensor element changes the incident light into current. Due to the stacked regions of the photo diode (Figure 3) 4 contacts are available. Based on the results of section 4 only 3 contacts has to be connected to the following circuitry. The foutrth contact a2 has to be multiplied by 0, so any further circuitry structure for this current can be neglected. Each of the other 3 contacts is connected to a preamplification stage, which is realized by current amplifiers. These stages implement the specific coefficients to the currents. At the summation point all currents were added. After this part the linear combination of the photo currents has finished. The following postamplification stage is only used to generate a vltage output signal. Therefor a transimpedance amplifier was used.

The integrated sensor circuitry was processed in a $0.35\,\mu$m-CMOS-technology. Afterwards measurements were executed to verify the ambient light sensor characteristic. The measurement setup was similar to the characterization of the simple stacked

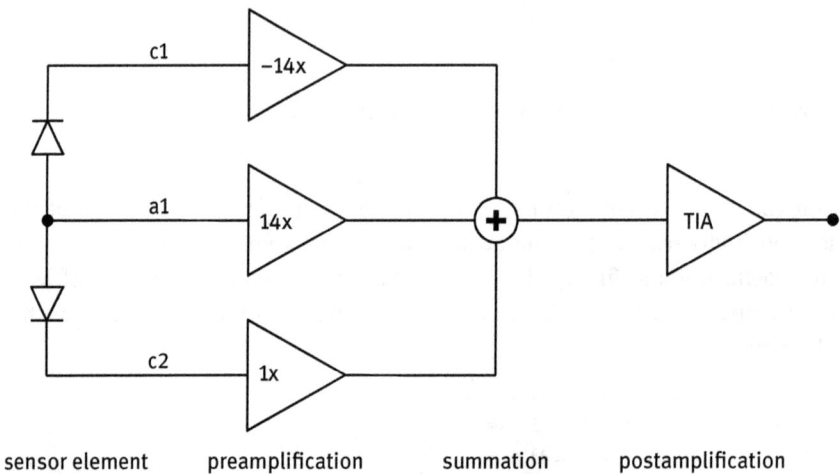

| sensor element | preamplification | summation | postamplification |

Fig. 5. Structure of the presented ambient light sensor.

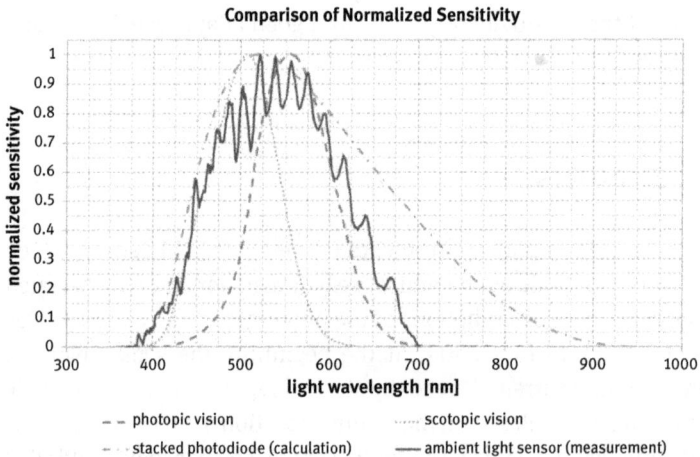

Fig. 6. Comparison of Normalized Sensitivity between calculation and measurement.

photo diode. Monochromatic light was applied to the sensor element and the resulting voltage was determied at the output. So the sensitivity characteristic could be recorded and compared to the human eye characteristic. The integrated ambient light sensor could not exactly meet the characteristic of photopic or scotopic vision, but it is a good approximation.

If a standard silicon photo diode is used as an ambient light sensor the relative sensitivity can change more than 90 % [11]. This parameter is defined as variation of sensitivity for different light sources in relation to standard light source A. The best commercial ambient light sensors can reach a deviation of less than 2%. The presented ambient light sensor shows a calculated variation of 11 % for incident sunlight.

6 Conclusion

An ambient light sensor can be efficiently realized by a stacked photodiode and an integrated circuitry. The sensor element is represented by a vertical stack of 3 pn-junctions within a single semiconductor device. The different sensitivity response for each pn-junction is caused due to variation of penetration depth for light wave lengths. Linear combination of each single resulting photo-generated current followed by a current summation allows adapting sensitivity response to the luminosity function of human eye. This functionality is given by the integrated circuitry. The sensor device was only realized by the usage of doped regions, which are parts of the existing $0.35\,\mu$m-CMOS-technology. Hence, no extra mask has to be applied to process the stacked photodiode. This realization saves processing time and costs. The strategy to adapt luminosity function can be reused to adapt sensitivity response to other required sensitivity functions. Therefore, a recalculation of the current multiplier coefficients is necessary.

Acknowledgment: This work was funded by the Thueringer Aufbaubank project number 2012FE9045 "EROLEDT – Energie- und ressourceneffizienter OLED-Treiber".

Bibliography

[1] H. Zimmermann. *Integrated Silicon Optoelectronics* Springer Verlag, Heidelberg, London, New York, Dortrecht, 2. edition, 2009.

[2] S.M. Sze. *Physics of Semiconductor devices* John Wiley and Sons New York, Chichester, Brisbane, Toronto, Singapore, 1. edition 1981.

[3] E. D. Palik. *Handbook of Optical Constants of Solids* Acadamic Press Boston, 1985.

[4] Y. Ohno. *OSA Handbook of optics, Volume III Optics and Vision Chapter for Photometry and Radiometry* Gaithersburg, USA, 1999.

[5]   CIE 018.2–1983. *The basis of Physical Photometry* 1983.
[6]   X-FAB Semiconductor Foundries AG *Process Specification XH035 0.35 μm Modular CMOS 2012*
      X-FAB Semiconductor Foundries AG, Release 5.0.1, 2012.
[7]   Prof. Dr. W. Schallreuter (ed.) *Lehrbuch der Physik* B. G. Teubner Verlagsgesellschaft, Leipzig,
      1969.
[8]   Synopsys Inc.. *Sentaurus Device User Guide* Version G2012–06, Mountain View, CA 94043,
      June 2012.
[9]   H. Zimmermann *Silicon Optoelectronic Integrated Circuits* Springer Verlag, Berlin Heidelberg,
      2004.
[10] L. Pavesi, D. J. Lockwood *Silicon Photonics* Springer Verlag Berlin Heidelberg, 2010.
[11] OSRAM GmbH *Ambient Light Sensors - General Application Note* OSRAM, 2006.

Biographies

Michael Meister received his diploma in electrical engineering from the Technical University in Ilmenau, Germany. In 2004 he joined IMMS and is responsible for optoelectronic and high temperature measurements. He is working on development of test setups for custom ASICs and characterization of integrated optoelectronic and high temperature devices.

Ulrich Liebold was born in 1955 and received the Diploma degree in electrical engineering with specialization on electron devices from the Ilmenau University of Technology in 1981. He is currently working as a device engineer and project manager for microelectronic and SOI high temperature application at the Institute for microelectronic and mechatronics systems. His research interests are TCAD device simulation of optoelectronic devices, especially behavior of photodiodes in the frequency domain.

Andrè Jäger was born in 1981. He received his diploma in electrical engineering and information technology from University of Applied Sciences Schmalkalden, Germany, in 2007. In 2007 he joined the Institute for Microelectronic and Mechatronic Systems GmbH, where he is currently working as design engineer and leading the analog IC design group.

Sebastian Thiele obtained his Master of Science (M.Sc.) degree in electrical engineering from the department of Solid State Electronics, TU Ilmenau. After his graduation in 2013 he joined X-FAB Semiconductor Foundries AG in Erfurt, Germany. His spectrum of tasks involves the characterization of integrated photo diodes and photo diode modelling.

Robin Weirauch received the diploma in physics from the department of Semiconductor Physics, University of Leipzig, Saxony, Germany, in 2007. He joined XFAB Semiconductor Foundries AG, Erfurt, in 2007, working in CMOS characterization in the field of device matching. He has also worked on Process Development for analog and high voltage applications.

Konrad Bach was a senior manager in X-FAB's process development department until 2014. He led the CMOS/BiCMOS development group based in Erfurt, Germany. Under his guidance, standard CMOS processes had been upgraded to complex modular process families that are focused on different mixed signal applications. These modular systems allow the customer to get off-the-shelf tailored solutions. He studied physics and gained his PhD in 1978 in the subject of Ion Implantation and its impact on minority carrier lifetime. Konrad Bach has more than 30 years of experience in industrial semiconductor process development. X-FAB's integrated High Voltage options and the integrated photo diode approach are strongly influenced by his technical expertise.

Daniel Gäbler is a principal engineer within the Analog and Sensors team at X-FAB and is based in Erfurt. He is working on the simulation, design and implementation of optical elements and sensors for integrating optical functionalities into foundry CMOS processes for more than 10 years. Daniel holds a diploma degree in Electrical Engineering from the Technical University in Ilmenau, Germany. He is author of 21 patents in the field of optical devices and X-FAB application expert for Sensors Ambient Light and Proximity Sensing.